Electricity and Magnetism Simulations
The Consortium for Upper-Level Physics Software

Robert Ehrlich and **Jaroslaw Tuszynski**
Department of Physics, George Mason University
Fairfax, Virginia

Lyle Roelofs
Department of Physics, Haverford College,
Pennsylvania

Ronald Stoner
Department of Physics, Bowling Green University
Bowling Green, Ohio

Series Editors

Maria Dworzecka

Robert Ehrlich

William MacDonald

JOHN WILEY & SONS, INC.

NEW YORK · CHICHESTER · BRISBANE · TORONTO · SINGAPORE

ACQUISITIONS EDITOR Cliff Mills
MARKETING MANAGER Catherine Faduska
SENIOR PRODUCTION EDITOR Sandra Russell
MANUFACTURING MANAGER Susan Stetzer

This book was set in 10/12 Times Roman by Beacon Graphics and
printed and bound by Hamilton Printing Co. The cover was printed by New England
Book Components, Inc.

Recognizing the importance of preserving what has been written, it is a
policy of John Wiley & Sons, Inc. to have books of enduring value published
in the United States printed on acid-free paper, and we exert our best
efforts to that end.

The paper on this book was manufactured by a mill whose forest management programs include
sustained yield harvesting of its timberlands. Sustained yield harvesting principles ensure that
the number of trees cut each year does not exceed the amount of new growth.

Library of Congress Cataloging in Publication Data:
Robert Ehrlich, Jaroslaw Tuszynski, Lyle Roelofs,
Ronald Stoner, Maria Dworzecka, William MacDonald
Electricity and Magnetism Simulations: The Consortium for Upper Level Physics Software

ISBN 0-471-54880-4 (pbk)

Printed in the United States of America

10 9 8 7 6 5 4 3 2 1

Contents

List of Figures

1

Introduction

"It is nice to know that the computer understands the problem. But I would like to understand it too."

—Eugene P. Wigner, quoted in *Physics Today,* July 1993

1.1   Using the Book and Software

The simulations in this book aim to exploit the capabilities of personal computers and provide instructors and students with valuable new opportunities to teach and learn physics, and help develop that all-important, if somewhat elusive, physical intuition. This book and the accompanying diskettes are intended to be used as supplementary materials for a junior- or senior-level course. Although you may find that you can run the programs without reading the text, the book is helpful for understanding the underlying physics, and provides numerous suggestions on ways to use the programs. *If you want a quick guided tour through the programs, consult the "Walk Throughs" in Appendix A.* The individual chapters and computer programs cover mainstream topics found in most textbooks. However, because the book is intended to be a supplementary text, no attempt has been made to cover all the topics one might encounter in a primary text.

Because of the book's organization, students or instructors may wish to deal with different chapters as they come up in the course, rather than reading the chapters in the order presented. One price of making the chapters semi-independent of one another is that they may not be entirely consistent in notation or tightly cross-referenced. Use of the book may vary according to the taste of the student or instructor. Students may use this material as the basis of a self-study course. Some instructors may make homework assignments from the large number of exercises in each chapter or to use them as the basis of student projects. Other instructors may use the computer programs primarily for in-class demonstrations. In this latter case, you may find that the programs are suitable for a range of courses from the introductory to the graduate level.

Use of the book and software may also vary with the degree of computer programming performed by users. For those without programming experience, all the computer simulations have been supplied in executable form, permitting them to be used as is. On the other hand, Pascal source code for the programs has also been provided, and a number of exercises suggest specific ways the programs can be modified. Possible modifications range from altering a single procedure especially set up for this purpose by the author, to larger modifications following given examples, to extensive additions for ambitious projects. However, the intent of the authors is that the simulations will help the student to develop intuition and a deeper understanding of the physics, rather than to develop computational skills.

We use the term "simulations" to refer to the computer programs described in the book. This term is meant to imply that programs include complex, often realistic, calculations of models of various physical systems, and the output is usually presented in the form of graphical (often animated) displays. Many of the simulations can produce numerical output—sometimes in the form of output files that could be analyzed by other programs. The user generally may vary many parameters of the system, and interact with it in other ways, so as to study its behavior in real time. The use of the term simulation should not convey the idea that the programs are bypassing the necessary physics calculations and simply producing images that look more or less like the real thing.

The programs accompanying this book can be used in a way that complements, rather than displaces, the analytical work in the course. It is our belief that, in general, computational and analytical approaches to physics can be mutually reinforcing. It may require considerable analytical work, for example, to modify the programs, or really to understand the results of a simulation. In fact, one important use of the simulations is to suggest conjectures that may then be verified, modified, or proven false analytically. A complete list of programs is given in Section 1.7.

1.2   Required Hardware and Installation of Programs

The programs described in this book have been written in the Pascal language for MS-DOS platforms. The language is Borland/Turbo Pascal, and the minimum hardware configuration is an IBM-compatible 386-level machine preferably with math coprocessor, mouse, and VGA color monitor. In order to accommodate a wide range of machine speeds, most programs that use animation include the capability to slow down or speed up the program. To install the programs, place disk number 1 in a floppy drive. Change to that drive, and type Install. You need only type in the file name to execute the program. Alternatively, you could type the name of the driver program (the same name as the directory in which the programs reside), and select programs from a menu. A number of programs write to temporary files, so you should check to see if your autoexec.bat file has a line that sets a temporary directory, such as SET TEMP = C:\TEMP. (If you have installed WINDOWS on your PC, you will find that such a command has already been written into your autoexec.bat file.) If no such line is there, you should add one.

Compilation of Programs

If you need to compile the programs, it would be preferable to do so using the Borland 7.0 (or later) compiler. If you use an earlier Turbo compiler you may run out of memory when compiling. If that happens, try compiling after turning off memory resident programs. If your machine has one, be sure to compile with the math-coprocessor turned on (no emulation). Finally, if you recompile programs using any compiler other than Borland 7.0, you will get the message: "EGA/VGA Invalid Driver File" when you try to execute them, because the driver file supplied was produced using this version of the compiler. In this case, search for the file BGILINK.pas included as part of the compiler to find information on how to create the EGAVGA.obj driver file. *If any other instructions are needed for installation, compilation, or running of the programs, they will be given in a README file on the diskettes.*

1.3   User Interface

To start a program, simply type the name of the individual or driver program, and an opening screen will appear. All the programs in this book have a common user interface. Both keyboard and mouse interactions with the computer are possible. Here are some conventions common to all the programs.

Menus: If using the *keyboard*, press **F10** to highlight one of menu boxes, then use the **arrow** keys, **Home**, and **End** to move around. When you press **Return** a submenu will pull down from the currently highlighted menu option. Use the same keys to move around in the submenu, and press **Return** to choose the highlighted submenu entry. Press **Esc** if you want to leave the menu without making any choices.

If using the *mouse* to access the top menu, click on the menu bar to pull down a submenu, and then on the option you want to choose. Click anywhere outside the menus if you want to leave them without making any choice. Throughout this book, the process of choosing submenu entry **Sub** under main menu entry **Main** is referred to by the phrase "choose **Main | Sub**." The detailed structure of the menu will vary from program to program, but all will contain **File** as the first (left-most) entry, and under **File** you will find **About CUPS, About Program, Configuration,** and **Exit Program**. The first two items when activated by mouse or arrows keys will produce information screens. Selecting **Exit Program** will cause the program to terminate, and choosing **Configuration** will present you with a list of choices (described later), concerning the mode of running the program. In addition to these four items under the **File** menu, some programs may have additional items, such as **Open**, used to open a file for input, and **Save**, used to save an output file. If **Open** is present and is chosen, you will be presented with a scrollable list of files in the current directory from which to choose.

Hot Keys: Hot keys, usually listed on a bar at the bottom of the screen, can be activated by pressing the indicated key or by clicking on the hot key bar with the mouse. The hot key **F1** is reserved for help, the hot key **F10** activates the menu bar. Other hot keys may be available.

Sliders (scroll-bars): If using the *keyboard*, press **arrow** keys for slow scrolling of the slider, **PgUp/PgDn** for fast scrolling, and **End/Home** for moving from one end to another. If you have more then one slider on the screen then only the slider with marked "thumb" (sliding part) will respond to the above keys. You can toggle the mark between your sliders by pressing the **Tab** key.

If using the *mouse* to adjust a slider, click on the thumb of the slider, drag it to desired value, and release. Click on the arrow on either end of the slider for slow scrolling, or in the area on either side of thumb for fast scrolling in this direction. Also, you can click on the box where the value of the slider is displayed, and simply type in the desired number.

Input Screens: All input screens have a set of "default" values entered for all parameters, so that you can, if you wish, run the program by using these original values. Input screens may include circular radio buttons and square check boxes, both of which can take on Boolean, i.e., "on" or "off," values. Normally, check boxes are used when only one can be chosen, and radio buttons when any number can be chosen.

If using the *keyboard*, press **Return** to accept the screen, or **Esc** to cancel it and lose the changes you may have made. To make changes on the input screen by keyboard, use **arrow** keys, **PgUp**, **PgDn**, **End**, **Home**, **Tab**, and **Shift-Tab** to choose the field you want to change, and use the backspace or delete keys to delete numbers. For Boolean fields, i.e., those that may assume one of two values, use any key except those listed above to change its value to the opposite value.

If you use the *mouse*, click [OK] to accept the screen or [Cancel] to cancel the screen and lose the changes. Use the mouse to choose the field you want to change. Clicking on the Boolean field automatically changes its value to the opposite value.

Parser: Many programs allow the user to enter expressions of one or more variables that are evaluated by the program. The function parser can recognize the following functions: absolute value (abs), exponential (exp), integer or fractional part of a real number (int or frac), real or imaginary part of a complex number (re or im), square or square root of a number (sqr or sqrt), logarithms—base 10 or e (log or ln)—unit step function (h), and the sign of a real number (sgn). It can also recognize the following trigonometric functions: sin, cos, tan, cot, sec, csc, and the inverse functions arcsin, arccos, arctan, as well as the hyperbolic functions denoted by adding an "h" at the end of all the preceding functions. In addition, the parser can recognize the constants pi, e, $i(\sqrt{-1})$, and rand (a random number between 0 and 1). The operations **+**, **−**, *****, **/**, **^**(exponentiation), and **!**(factorial) can all be used, and the variables r and c are interpreted as $r = \sqrt{x^2 + y^2}$ and $c = x + iy$. Expressions involving these functions, variables, and constants can be nested to an arbitrary level using parentheses and brackets. For example, suppose you entered the following expression: **h(abs(sin(10*pi* x))−0.5)**. The parser would interpret this function as $h(|sin(10\pi x)|-0.5)$. If the program evaluates this function for a range of x-values, the result, in this case, would be a series of square pulses of width 1/15, and center-to-center separation 1/10.

Help: Most programs have context-sensitive help available by pressing the **F1** hot key (or clicking the mouse in the **F1** hot key bar). In some programs help is also available by choosing appropriate items on the menu, and in still other programs tutorials on various aspects of the program are available.

1.4   The CUPS Project and CUPS Utilities

The authors of this book have developed their programs and text as part of the Consortium for Upper-Level Physics Software (CUPS). Under the direction of the three editors of this book, CUPS is developing computer simulations and associated texts for nine junior- or senior-level courses, which comprise most of the undergraduate physics major curriculum during those two years. A list of the nine CUPS courses, and the authors associated with each course, follows this section. This international group of 27 physicists includes individuals with extensive backgrounds in research, teaching, and development of instructional software.

The fact that each chapter of the book has been written by a different author means that the chapters will reflect that individual's style and philosophy. Every attempt has been made by the editors to enhance the similarity of chapters, and to provide a similar user interface in each of the associated computer simulations. Consequently, you will find that the programs described in this and other CUPS books have a common look and feel. This degree of similarity was made possible by producing the software in a large group that shared a common philosophy and commitment to excellence.

Another crucial factor in developing a degree of similarity between all CUPS programs is the use of a common set of utilities. These CUPS utilities were written by Jaroslaw Tuszynski and William MacDonald, the former having responsibility for the graphics units, and the latter for the numerical procedures and functions. The numerical algorithms are of high quality and precision, as required for reliable results. CUPS utilities were originally based on the M.U.P.P.E.T. utilities of Jack Wilson and E.F. Redish, which provided a framework for a much expanded and enhanced mathematical and graphics library. The CUPS utilities (whose source code is included with the simulations with this book), include additional object-oriented programs for a complete graphical user interface, including pull-down menus, sliders, buttons, hotkeys, and mouse clicking and dragging. They also include routines for creating contour, two-dimensional (2-D) and 3-D plots, and a function parser. The CUPS utilities have been provided in source code form to enable users to run the simulations under future generations of Borland/Turbo Pascal. If you do run under future generations of Turbo or Borland Pascal on the PC, the utilities and programs will need to be recompiled. You will also need to create a new egavga.obj file which gets combined with the programs when an executable version is created—thereby avoiding the need to have separate (egavga.bgi) driver files. These CUPS utilities are also available to users who wish to use them for their own projects.

One element not included in the utilities is a procedure for creating hard copy based on screen images. When hard copy is desired, those PC users with the appropriate graphics driver (graphics.com), may be able to produce high-quality screen images by depressing the **PrintScreen** key. If you do not have the graphics software installed to get screen dumps, select **Configuration | Print Screen**, and follow the directions. Moreover, public domain software also exists for capturing screen images, and for producing PostScript files, but the user should be aware that such files are often quite large, sometimes over 1 MB, and they require a PostScript printer driver to produce.

One feature of the CUPS utilities that can improve the quality of hard copy produced from screen captures is a procedure for switching colors. This capability is important because the gray scale rendering of colors on black-and-white printers may create poor contrasts if the original (default) color assignments are used. To access the CUPS utility for changing colors, the user need only choose **Configuration** under the **File** menu when the program is first initiated, or at any later time. Once you have chosen **Configuration**, to change colors you need to click the mouse on the **Change Colors** bar, and you will be presented with a 16 by 16 matrix of radio buttons that will allow you to change any color to any other color, or else to use predefined color switches, such as a color "reversal," or a conversion of all light colors to black, and all dark colors to white. (The screen captures given in this book were produced using the "reverse" color map.) Any such color changes must be redone when the program is restarted.

Other system parameters may likewise be set from the **File | Configuration** menu item. These include the path for temporary files that the program may create (or want to read), the mouse "double click" speed—important for those with slow reflexes—an added time delay to slow down programs on computers that are too fast, and a "check memory" option—primarily of interest to those making program modifications.

Those users wishing more information on the CUPS utilities should consult the CUPS Utilities Manual, written by Jaroslaw Tuszynski and William Mac-Donald, published by John Wiley and Sons. However, it is not necessary for casual users of CUPS programs to become familiar with the utilities. Such familiarity would only be important to someone wishing to write their own simulations using the utilities. The utilities are freely available for this purpose, for unrestricted noncommercial production and distribution of programs. However, users of the utilities who wish to write programs for commercial distribution should contact John Wiley and Sons.

1.5   *Communicating With the Authors*

Users of these programs should not expect that run-time errors will never occur! In most cases, such run-time errors may require only that the user restart the program; but in other cases, it may be necessary to reboot the computer, or even turn it off and on. The causes of such run-time errors are highly varied. In some cases, the program may be telling you something important about the physics or the numerical method. For example, you may be trying to use a numerical method beyond its range of applicability. Other types of run-time errors may have to do with memory or other limitations of your computer. Finally, although the programs in this book have been extensively tested, we cannot rule out the possibility that they may contain errors. (Please let us know if you find any! It would be most helpful if such problems were communicated by electronic mail, and with complete specificity as to the circumstances under which they arise.)

It would be best if you communicated such problems directly to the author of each program, and simultaneously to the editors of this book (the CUPS Directors), via electronic mail—see addresses listed below. Please feel free to communicate any suggestions about the programs and text which may lead to improvements in future editions. Since the programs have been provided in source code form, it will be possible for you to make corrections of any errors that you or we find in the future—provided that you send in the registration card at the back of the book, so that you can be notified. The fact that you have the source code will also allow you to make modifications and extensions of the programs. We can assume no responsibility for errors that arise in programs that you have modified. In fact, we strongly urge you to change the program name, and to add a documentary note at the beginning of the code of any modified programs that alerts other potential users of any such changes.

1.6   CUPS Courses and Developers

- **CUPS Directors**
  Maria Dworzecka, George Mason University (cups@gmuvax.gmu.edu)
  Robert Ehrlich, George Mason University (cups@gmuvax.gmu.edu)
  William MacDonald, University of Maryland (w_macdonald@umail.umd.edu)

- **Astrophysics**
  J. M. Anthony Danby, North Carolina State University (n38hs901@ncuvm.ncsu.edu)
  Richard Kouzes, Battelle Pacific Northwest Laboratory (rt_kouzes@pnl.gov)
  Charles Whitney, Harvard University (whitney@cfa.harvard.edu)

- **Classical Mechanics**
  Bruce Hawkins, Smith College (bhawkins@smith.bitnet)
  Randall Jones, Loyola College (rsj@loyvax.bitnet)

- **Electricity and Magnetism**
  Robert Ehrlich, George Mason University (rehrlich@gmuvax.gmu.edu)
  Lyle Roelofs, Haverford College (lroelofs@haverford.edu)
  Ronald Stoner, Bowling Green University (stoner@andy.bgsu.edu)
  Jaroslaw Tuszynski, George Mason University (cups@gmuvax.gmu.edu)

- **Modern Physics**
  Douglas Brandt, Eastern Illinois University (cfdeb@ux1.cts.eiu.edu)
  John Hiller, University of Minnesota, Duluth (jhiller@d.umn.edu)
  Michael Moloney, Rose Hulman Institute (moloney@nextwork.rose-hulman.edu)

- **Nuclear and Particle Physics**
  Roberta Bigelow, Willamette University (rbigelow@willamette.edu)
  John Philpott, Florida State University (philpott@fsunuc.physics.fsu.edu)
  Joseph Rothberg, University of Washington (rothberg@phast.phys.washington.edu)

- **Quantum Mechanics**
  John Hiller, University of Minnesota, Duluth (jhiller@d.umn.edu)
  Ian Johnston, University of Sydney (idj@suphys.physics.su.oz.au)
  Daniel Styer, Oberlin College (dstyer@physics.oberlin.edu)

- **Solid State Physics**
  Graham Keeler, University of Salford (g.j.keeler@sysb.salford.ac.uk)
  Roger Rollins, Ohio University (rollins@chaos.phy.ohiou.edu)
  Steven Spicklemire, University of Indianapolis (steves@truevision.com)

- **Thermal and Statistical Physics**
  Harvey Gould, Clark University (hgould@vax.clarku.edu)
  Lynna Spornick, Johns Hopkins University
  Jan Tobochnik, Kalamazoo College (jant@kzoo.edu)

- **Waves and Optics**
  G. Andrew Antonelli, Wolfgang Christian, and Susan Fischer, Davidson College (wc@phyhost.davidson.edu)
  Robin Giles, Brandon University (giles@brandonu.ca)
  Brian James, Salford University (b.w.james@sysb.salford.ac.uk)

1.7   Descriptions of all CUPS Programs

Each of the computer simulations in this book (as well as those in the eight other books comprised by the CUPS Project) are described below. The individual headings under which programs appear correspond to the nine CUPS courses. In several cases, programs are listed under more than one course. The number of programs listed under the Astrophysics, Modern Physics, and Thermal Physics courses is appreciably greater than the others, because several authors have opted to subdivide their programs into many smaller programs. Detailed inquiries regarding CUPS programs should be sent to the program authors.

ASTROPHYSICS PROGRAMS

STELLAR (Stellar Models), written by Richard Kouzes, is a simulation of the structure of a static star in hydrodynamic equilibrium. This provides a model of a zero age main sequence star, and helps the user understand the physical processes that exist in stars, including how density, temperature, and luminosity depend on mass. Stars are self-gravitating masses of hot gas supported by thermodynamic processes fueled by nuclear fusion at their core. The model integrates the four differential equations governing the physics of the star to reach an equilibrium condition which depends only on the star's mass and composition.

EVOLVE (Stellar Evolution), written by Richard Kouzes, builds on the physics of a static star, and considers (1) how a gas cloud collapses to become a main sequence star, and (2) how a star evolves from the main sequence to its final demise. The model is based on the same physics as the STELLAR program. Starting from a diffuse cloud of gas, a protostar forms as the cloud collapses and reaches a sufficient density for fusion to begin. Once a star reaches equilibrium, it remains for most of its life on the main sequence, evolving off after it has consumed its fuel. The final stages of the star's life are marked by rapid and dramatic evolution.

BINARIES is the driver program for all Binaries programs (**VISUAL1, VISUAL2, ECLIPSE, SPECTRO, TIDAL, ROCHE, and ACCRDISK**).

VISUAL1 (Visual Binaries—Proper Motion), written by Anthony Danby, enables you to visualize the proper motion in the sky of the members of a visual binary system. You can enter the elements of the system and the mass ratio, as well as the speed at which the center of mass moves across the screen. The program also includes an animated three-dimensional demonstration of the elements.

VISUAL2 (Visual Binaries—True Orbit), written by Anthony Danby, enables you to select an apparent orbit for the secondary star with arbitrary eccentricity, with the primary at any interior point. The elements of the orbit are displayed. You can see the orbit animated in three dimensions, or can make up a set of "observations" based on the apparent orbit.

ECLIPSE (Eclipsing Binaries), written by Anthony Danby, shows simultaneously either the light curve and the orbital motion or the light curve and an animation of the eclipses. You can select the elements of the orbit and radii and magnitudes of the stars. A form of limb-darkening is also included as an option.

SPECTRO (Spectroscopic Binaries), written by Anthony Danby, allows you to select the orbital elements of a spectroscopic binary, and then shows simultaneously the velocity curve, the orbital motion, and a moving spectral line.

TIDAL (Tidal Distortion of a Binary), written by Anthony Danby, models the motion of a spherical secondary star around a primary that is tidally distorted by the secondary. You can select orbital elements, masses of the stars, a parameter describing the tidal lag, and the initial rate of rotation of the primary. The equations are integrated over a time interval that you specify. Then you can see the changes of the orbital elements, and the rotation of the primary, with time. You can follow the motion in detail around each revolution, or in a form where the equations have been averaged around each revolution.

ROCHE (The Photo-Gravitational Restricted Problem of Three Bodies), written by Anthony Danby, follows the two-dimensional motion of a particle that is subject to the gravitational attraction of two bodies in mutual circular orbits, and also, optionally, radiation pressure from these bodies. It is intended, in part, as background for the interpretation of the formation of accretion disks. Curves of zero velocity (that limit regions of possible motion) can be seen. The orbits can also be followed using Poincaré maps.

ACCRDISK (Formation of an Accretion Disk), written by Anthony Danby, follows some of the dynamical steps in this process. The dynamics is valid up to the initial formation of a hot spot, and qualititative afterward.

NBMENU is the driver program for all programs on the motion of N interacting bodies: **TWO-GALAX, ASTROIDS, N-BODIES, PLANETS, PLAYBACK, and ELEMENTS**.

TWOGALAX (The Model of Wright and Toomres), written by Anthony Danby, is concerned with the interaction of two galaxies. Each consists of a central gravitationally attracting point, surrounded by rings of stars (which are attracted, but do not attract). Elements of the orbits of one galaxy relative to the other are selected, as is the initial distribution and population of the rings. The motion can be viewed as projected into the plane of the orbit of the galaxies, or simultaneously in that plane and perpendicular to it. The positions can be stored in a file for later viewing.

ASTROIDS (N-Body Application to the Asteroids), written by Anthony Danby, uses the same basic model, but a planet and a star take the place of the galaxies and the asteroids replace the stars. Emphasis is on asteroids all having the same period, with interest on periods having commensurability with the period of the planet. The orbital motion of the system can be followed. The positions can be stored in a file for later viewing. An asteroid can be selected, and the variation of its orbital elements can then be followed.

NBODIES (The Motion of N Attracting Bodies), written by Anthony Danby, allows you to choose the number of bodies (up to 20) and the total energy of the system. Initial conditions are chosen at random, consistent with this energy, and the resulting motion can be observed. During the motion various quantities, such as the kinetic energy, are displayed. The positions can be stored in a file for later viewing.

PLANETS (Make Your Own Solar System), written by Anthony Danby, is similar to the preceding program, but with the bodies interpreted as a star with planets. Initial conditions are specified through the choice of the initial elements of the planets. The positions can be stored in a file for later viewing.

PLAYBACK, written by Anthony Danby, enables a file stored by one of the preceding programs to be viewed.

ELEMENTS (Orbital Elements of a Planet), written by Anthony Danby, shows a three-dimensional animation that can be viewed from any angle.

GALAXIES is the driver program for Galactic Kinematics programs: **ROTATION, OORTCONS, and ARMS21CM**.

ROTATION (The Rotation Curve of a Galaxy), written by Anthony Danby, first prompts you to "design" a galaxy, consisting of a central mass and up to five spheroids (that can be visible or invisible). It then displays the galaxy and can show the animated rotation or the rotation curve.

OORTCONS (Galactic Kinematics and Oort's Constants), written by Anthony Danby, allows you to design your galaxy, choose the location of the "sun" and a local region around it, and the to observe the kinematics in this region. It also shows graphs of radial velocity and proper motion in comparison with the linear approximation, and computes the Oort constants.

ARMS21CM (The Spiral Structure of a Galaxy), written by Anthony Danby, allows you to design your galaxy, construct a set of spiral arms, and select the position of the "sun." Then, for different galactic longitudes, you can see observed profiles of 21 cm lines.

ATMOS (Stellar Atmospheres), written by Charles Whitney, permits the user to select a constellation, see it mapped on the computer screen, point to a star, and see it plotted on a brightness-color diagram. The user's task is to build a model atmosphere that imitates the photometric properties of observed stars. This is done by specifying numerical values for three basic stellar parameters: radius, mass, and luminosity. The program then builds the model and displays it on the brightness-color diagram, and it also plots the spectrum and the detailed thermodynamic structure of the atmosphere. With this program the user may investigate the relation between stellar parameters and the thermal properties of the gas in the atmosphere. Two atmospheres may be superposed on the graphs, for easier comparison.

PULSE (Stellar Pulsations), written by Charles Whitney, illustrates stellar pulsation by simulating the thermo-mechanical behavior of a "star" modeled by a self-gravitating gas divided by spherical elastic shells. The elastic shells resemble a set of coupled oscillators. The program solves for the modes of small-amplitude motion, and it uses Fourier synthesis to construct motions for arbitrary starting conditions. The screen displays the thermodynamic structure and surface properties, such as temperature, pressure, and velocity. Animation displays the nature of the pulsation. By showing the motions, temperatures, and energy flux, the program demonstrates the heat engine acting inside the pulsating star. The motions of the shells and the spatial Fourier decomposition into eigenmodes are displayed simultaneously, and this will help you visualize the meaning of the Fourier components.

CLASSICAL MECHANICS PROGRAMS

GENMOT (The Motion Generator), written by Randall Jones, allows you to solve numerically any differential equation of motion for a system with up to three degrees of freedom and display the time evolution of the system in a wide variety of formats. Any of the dynamical variables or any function of those variables may be displayed graphically and/or numerically and a wide range of animations may be constructed. Since the Motion Generator can be used to solve any second-order differential equation, it can also be used to study systems analyzed by Lagrangian methods. Real world coordinates may be constructed as functions of generalized coordinates so that simulations of the actual system can be constructed.

ROTATE (Rotation of 3-D Objects), written by Randall Jones, is designed to aid in the visualization of the dynamical variables of rotational motion. It will allow you to observe the 3-D motion of rotating objects in a controlled fashion, running the simulation faster, slower, or in reverse while displaying the corresponding evolution of the angular velocity, the angular momentum and the torque. It will display the motion from the fixed frame and from the body frame to help in understanding the translation between these two descriptions of the motion. By using the stereographic feature of the program you can create a genuine 3-D representation of the motion of the quantities.

COUPOSC (Coupled Oscillators), written by Randall Jones, is designed to investigate a wide range of harmonic systems. Given a set of objects and springs connected in one or two dimensions, the simulation can solve the problem by generating the normal mode frequencies and their corresponding motions. It can take any set of initial conditions and resolve them into their component normal mode motions or take any set of initial mode occupations and display the corresponding motions of the objects. It can also determine the motion of the system when it is acted on by external forces. In this case the total forces are no longer harmonic, so the solution is generated numerically. The harmonic analysis, however, still provides an important tool for investigating and understanding the subsequent motion.

ANHARM (Anharmonic Oscillators), written by Bruce Hawkins, simulates oscillations of various types: pendulum, simple harmonic oscillator, asymmetric, cubic, Vanderpol, and a mass in the center of a spring with fixed ends. Nonlinear behavior is emphasized. The user may choose to view one to four graphs of the motion simultaneously, along with the potential diagram and a picture of the moving object. Graphs that may be viewed are x vs. t, v vs. t, v vs. x, the Poincaré diagram, and the return map. Tools are provided to explore parameter space for regions of interest. Fourier analysis is available, resonance diagrams can be plotted, and the period can be plotted as a function of energy. Includes a tutorial demonstrating the usefulness of phase plots and Poincaré plots.

ORBITER (Gravitational Orbits), written by Bruce Hawkins, simulates the motion of up to five objects with mutually gravitational attraction, and any reasonable number of additional objects moving in the gravitation field of the first five. The motion may be viewed in up to six windows simultaneously: windows centered on a particular body, on the center of mass, stationary in the universe frame, or rotating with the line joining the two most massive bodies. A menu of available systems includes the solar system, the sun/earth/moon system; the sun, Jupiter, and its moons; the sun, earth, and Saturn, demonstrating retrograde motion; the sun, Jupiter, and a comet; and a pair of binary stars with a comet. Bodies may be added to any system, or a new system created using either numerical coordinates or the mouse. Bodies may be replicated to demonstrate the sensitivity of orbits to initial conditions.

COLISION (Collisions), written by Bruce Hawkins, simulates two-body collisions under any of a number of force laws: Coulomb with and without shielding and truncation, hard sphere, soft sphere (harmonic), Yukawa, and Woods-Saxon. Collision may be viewed in the laboratory and center of mass systems simultaneously, with or without momentum diagrams. Includes a tutorial on the usefulness of the center of mass system, one on the kinematics of relativistic collisions, and one on cross section. Plots cross section against scattering parameter, and compares collisions at different parameters.

ELECTRICITY AND MAGNETISM PROGRAMS

FIELDS (Analysis of Vector and Scalar Fields), written by Jarek Tuszynski, displays scalar and vector fields for any algebraic or trigonometric expression entered by the user. It also computes numerically the divergence, curl, and Laplacian for the vector fields, and the gradient and Laplacian for the scalar fields. Simultaneous displays of selected quantities are provided in user-selected planes, using vector, contour, or 3-D plots. The program also allows the user to define paths along which line integrals are computed.

GAUSS (Gauss' Law), written by Jarek Tuszynski, treats continuous charge distributions having spherical or cylindrical symmetry, and those that vary as a function of the x-coordinate only. The program allows the user to enter an arbitrary function to define either the electric field magnitude, the potential, or the charge density. It then computes the other two functions by numerical differentiation or integration, and displays all three functions. Finally, the program allows the user to enter a "comparison function," which is plotted on the same graph, so as to check whether his analytic solutions are correct.

POISSON (Poisson's Equation Solved on a Grid), written by Jarek Tuszynski, solves Poisson's equation iteratively on a 2-D grid using the method of simultaneous over-relaxation. The user can draw arbitrary systems consisting of line charges, and charged conducting cylinders, plates, and wires, all infinite in extent perpendicular to the grid. After iteratively solving Poisson's equation, the program displays the results for the potential, electric field, or the charge density (found from the Laplacian of the potential), in the form of contour, vector, or 3-D plots. In addition, many other program features are available, including the ability to specify surfaces, along which the potential varies according to some algebraic function specified by the user.

IMAG&MUL (Image Charges and Multipole Expansion), written by Lyle Roelofs and Nathaniel Johnson, allows users to explore two approaches to the solution of Laplace's equation—the image charge method and expansion in multipole moments. In the image charge mode (IC) the user is presented with a variety of configurations involving conducting planes and point charges and is asked to "solve" each by placing image charges in the appropriate locations. The program displays the electric field due to all point charges, real and image, and a solution can be regarded as successful with the field due to all charges is everywhere orthogonal to all conducting surfaces. Solutions can then be examined with a variety of included software "tools." The multipole expansion (ME) mode of the program also permits a "hands-on" exploration of standard electrostatic problems, in this case the "exterior" problem, i.e., the determination of the field outside a specified equipotential surface. The program presents the user with a variety of azimuthally symmetric equipotential surfaces. The user "solves" for the full potential by adding chosen amounts of the (first six) multipole moments. The screen shows the contours of the summed potential and the problem is "solved" when the innermost contour matches the given equipotential surface as closely as possible.

ATOMPOL (Atomic Polarization), written by Lyle Roelofs and Nathaniel Johnson, is an exploration of the phenomenon of atomic polarization. Up to 36 atoms of controllable polarizability are immersed in an external electric field. The program solves for and displays the field throughout the region in which the atoms are located. A closeup window shows the polarization of selected atoms and software "tools" allow for further analysis of the resulting electric fields. Use of this program improves the student's understanding of polarization, the interaction of polarized entities, and the atomic origin of macroscopic polarization, the latter via study of closely spaced clusters of polarizable atoms.

DIELECT (Dielectric Materials), written by Lyle Roelofs and Nathaniel Johnson, is a simulation of the behavior of linear dielectric materials using a cell-based approach. The user controls either the polarization or the susceptibility of each cell in a (25×25) grid (with assumed uniformity in the third direction). Full self-consistent solutions are obtained via an iterative relaxation method and the fields P, E, or D are displayed. The student can investigate the self-interactions of polarized materials and many geometrical effects. Use of this program aids the student in developing understanding of the subtle relations among and meaning of P, E, and D.

ACCELQ (Fields From an Accelerated Charge), written by Ronald Stoner, simulates the electromagnetic fields in the plane of motion generated by a point charge that is moving and accelerating in two dimensions. The user chooses from among seven predefined trajectories, and sets the values of maximum speed and viewing time. The electric field pattern is recomputed after each change of trajectory or parameter; thereafter, the user can investigate the electric field, magnetic field, retarded potentials, and Poynting-vector field by using the mouse as a field probe, by using gridded overlays, or by generating plots of the various fields along cuts through the viewing plane.

QANIMATE (Fields From an Accelerated Charge—Animated Version), written by Ronald Stoner, is an interactive animation of the changing electric field pattern generated by a point electric charge moving in two dimensions. Charge motion can be manipulated by the user from the keyboard. The display can include electric field lines, radiation wave fronts, and their points of intersection. The motion of the charge is controlled by the using **arrow** keys to accelerate and steer much like the accelerator and steering wheel of a car, except that acceleration must be changed in increments, and the **Space** bar can used to engage or disengage the steering. With steering engaged, the charge will move in a circle. Unless the acceleration is made zero, the speed will increase (or decrease) to the maximum (minimum) possible value. At constant speed and turning rate, the charge can be controlled by the **Space** bar alone.

EMWAVE (Electromagnetic Waves), written by Ronald Stoner, uses animation to illustrate the behavior of electric and magnetic fields in a polarized plane electromagnetic wave. The user can choose to observe the wave in free space, or to see the effect on the wave of incidence on a material interface, or to see the effects of optical elements that change its polarization. The user can change the polarization state of the incident wave by specifying its Stokes parameters. Standing electromagnetic waves can be simulated by combining the incident traveling wave with a reflected wave of the same amplitude. The user can do that by choosing appropriate values of the physical properties of the medium on which the incident wave impinges in one of the animations.

MAGSTAT (Magnetostatics), written by Ronald Stoner, computes and displays magnetic fields in and near magnetized materials. The materials are uniform and have 3-D shapes that are solids of revolution about a vertical axis. The shape of the material can be modified or chosen from a data input screen. The user has the option of generating the fields produced by a permanently and uniformly magnetized object, or of generating the fields of a magnetizable object placed in an otherwise uniform external field. Besides choosing the shape and aspect ratio of the object, the user can vary the magnetic permeability of the magnetizable material, and choose among three fields to display: magnetic induction (B), magnetic field strength (H), and magnetization (M). Each of these fields can be displayed or explored in several different ways. The algorithm for computing the fields uses a superposition of Chebyschev polynomial approximants to the H field due to "rings" of "magnetic charge."

MODERN PHYSICS PROGRAMS

NUCLEAR (Nuclear Energetics and Nuclear Counting), written by Michael Moloney, deals with basic nuclear properties related to mass, charge, and energy, for approximately 1900 nuclides. Graphs are available involving binding energy, mass, and Q values of a variety of nuclear reactions, including alpha and beta decays. Part 2 deals with simulating the statistics of counting with a Geiger-Muller tube. This part also simulates neutron activation, and the counting behavior as neutron flux is turned on and off. Finally, a decay chain from A to B to C is simulated, where half-lives may be changed, and populations are graphed as a function of time.

GERMER (Davisson-Germer and G. P. Thomson Experiments), written by Michael Moloney, simulates both the Davisson-Germer and G. P. Thomson experiments with electrons scattering from crystalline materials. Stress is laid on the behavior of electrons as waves; similarities are noted with scattering of x-rays. The exercises encourage students to understand why peaks and valleys in scattered electrons occur where they do.

QUANTUM (one-dimensional Quantum Mechanics), written by Douglas Brandt, is a program that has four sections. The first section allows users to investigate the uncertainty principle for specified wavefunctions in position or momentum space. The second section allows users to investigate the time evolution of wavepackets under various dispersion relations. The third section allows users to investigate solutions to Schrödinger's equation for asymptotically free solutions. The user can input a barrier and the program calculates reflection and transmission coefficients for a range of energies and show wavepacket time evolution for the barrier potential. The fourth section is similar to the third, except that it allows the user to investigate bound solutions to Schrödinger's equation. The program calculates the bound state Hamiltonian eigenvalues and spatial eigenfunctions.

RUTHERFD (Rutherford Scattering), written by Douglas Brandt, is a program for investigating classical scattering of particles. A scattering potential can be chosen from a list of predefined potentials or an arbitrary potential can be input by the user. The computer generates scattering events by randomly picking impact parameters from a distribution defined by beam parameters specified by the user. It displays the results of the scattering on a polar histogram and on a detailed histogram to help users gain insight into differential scattering cross section. A scintillation mode can be chosen for users that want more appreciation of the actual experiments of Geiger and Marsden. A "guess the scatterer" mode is available for trying to gain appreciation of how scattering experiments are used to infer properties of the scatterers.

SPECREL (Special Relativity), written by Douglas Brandt, is a program to investigate special relativity. The first section is to investigate change of coordinate systems through Minkowski diagrams. The user can define coordinates of objects in one reference frame and the computer calculates the coordinates in a user-selectable coordinate system and displays the objects in both reference frames. The second section allows users to view clocks that are in relative motion. A clock can be given an arbitrary trajectory through space-time and the readings of various clocks can be viewed as the clock follows that trajectory. A third section allows users to observe collisions in different reference frames that are related by Lorentz transformations or by Gallilean transformations.

LASER (Lasers), written by Michael Moloney, simulates a three-level laser, with the user in control of energy level parameters, temperature, pump power, and end mirror transmission. Atomic populations may be graphically tracked from thermal equilibrium through the lasing threshold. A mirror cavity simulation is available which uses ray tracing. This permits study of cavity stability as a function of mirror shape and position, as well as beam shape characteristics within the cavity.

HATOM (Hydrogenic Atoms), written by John Hiller, computes eigenfunctions and eigenenergies for hydrogen, hydrogenic atoms, and single-electron diatomic ions. Hydrogenic atoms may be exposed to uniform electric and magnetic fields. Spin interactions are not included. The magnetic interaction used is the quadratic Zeeman term; in the absence of spin-orbit coupling, the linear term adds only a trivial energy shift. The unperturbed hydrogenic eigenfunctions are computed directly from the known solutions. When external fields are included, approximate results are obtained from basis-function expansions or from Lanczos diagonalization. In the diatomic case, an effective nuclear potential is recorded for use in calculation of the nuclear binding energy.

NUCLEAR AND PARTICLE PHYSICS PROGRAMS

NUCLEAR (Nuclear Energetics and Counting), written by Michael Moloney, is included here, but is described under the Modern Physics heading.

SHELLMOD (Nuclear Models), written by Roberta Bigelow, calculates energy levels for spherical and deformed nuclei using the single particle shell model. You can explore how the nuclear potential shape, the spin-orbit interaction, and deformation affect both the order and spacing of nuclear energy levels. In addition, you will learn how to predict spin and parity for single particle states.

NUCRAD (Interaction of Radiation With Matter), written by Roberta Bigelow, is a simulation of alpha particles, muons, electrons, or photons interacting with matter. You will develop an understanding of how ranges, energy losses, and random particle paths depend on materials, radiation, and incident energy. As a specific application, you can explore photon and electron interactions in a sodium iodide crystal which determines the energy response of a radiation detector.

ELSCATT (Electron-Nucleus Scattering), by John Philpott, is an interactive software tool that demonstrates various aspects of electron scattering from nuclei. Specific features include the relativistic kinematics of electron scattering, densities and form factors for elastic and inelastic scattering, and the nuclear Coulomb response. The simulation illustrates how detailed nuclear structure information can be obtained from electron scattering measurements.

TWOBODY (Two-Nucleon Interactions), by John Philpott, is an interactive software tool that illuminates many features of the two-nucleon problem. Bound state wavefunctions and properties can be calculated for a variety of interactions that may include non-central parts. Phase shifts and cross sections for pp, pn, and nn scattering can be calculated and compared with those obtained experimentally. Spin-polarization features of the cross sections can be extensively investigated. The simulation demonstrates the richness of the two-nucleon data and its relation to the underlying nucleon-nucleon interaction.

RELKIN (Relativistic Kinematics), by Joseph Rothberg, is an interactive program to permit you to explore the relativistic kinematics of scattering reactions and two-body particle decays. You may choose from among a large number of initial and final states. The initial momentum of the beam particle and the center of mass angle of a secondary can also be specified. The program displays the final state vector momenta in both the lab system and center of mass system along with numerical values of the most important kinematic quantities. The program may be run in a Monte Carlo mode, displaying a scatter plot and histogram of selected variables. The particle data base may be modified by the user and additional reactions and decay modes may be added.

DETSIM (Particle Detector Simulation), by Joseph Rothberg, is an interactive tool to allow you to explore methods of determining parameters of a decaying particle or scattering reaction. The program simulates the response of high-energy particle detectors to the final-state particles from scattering or decays. The detector size and location may be specified by the user as well as its energy and spatial resolution. If the program is run in a Monte Carlo mode, detector hit information for each event is written to a file. This file can be read by a small reconstruction and plotting program. You can easily modify one of the example reconstruction programs that are provided to determine the mass, momentum, and other properties of the initial particle or state.

QUANTUM MECHANICS PROGRAMS

BOUND1D (Bound States in One Dimension), written by Ian Johnston, is a tool which allows you to explore energy eigenfunctions for an electron in various potential wells, which can be square, parabolic, ramped, asymmetric, double, or Coulombic. The first part of the program deals with finding the eigenvalues and eigenfunctions of different wells. You may find them yourself, using a "hunt and shoot" method, or else the program will compute the eigenvalues automatically, by counting the number of nodes to determine where the eigenvalues occur. The second part of the program looks at properties of eigenfunctions normalization, orthogonality, and the evaluation of many kinds of overlap integrals. The third part examines the time development of general states made up of a superposition of bound state eigenfunctions. Facility is provided for you to incorporate your own procedures to specify different potential wells or different overlap integrals.

SCATTR1D (Scattering in One Dimension), written by John Hiller, solves the time-independent Schrödinger equation for stationary scattering states in one-dimensional potentials. The wavefunction is displayed in a variety of ways, and the transmission and reflection probabilities are computed. The probabilities may be displayed as functions of energy. The computations are done by numerically integrating the Schrödinger equation from the region of the transmitted wave, where the wavefunction is known up to some overall normalization and phase, to the region of the incident wave. There the reflected and incident waves are separated. The potential is assumed to be zero in the incident region and constant in the transmitted region.

QMTIME (Quantum Mechanical Time Development), written by Daniel Styer, simulates quantal time development in one dimension. A variety of initial wave packets (Gaussian, Lorentzian, etc.) can evolve in time under the influence of a variety of potential energy functions (step, ramp, square well, harmonic oscillator, etc.) with or without an external driving force. A novel visualization technique simultaneously displays the magnitude and phase of complex-valued wave functions. Either position-space or momentum-space wave functions, or both, can be shown. The program is particularly effective in demonstrating the classical limit of quantum mechanics.

LATCE1D (Wavefunctions on a one-dimensional Lattice), written by Ian Johnston, is a tool which allows you to explore energy eigenfunctions for an electron in a lattice made up of a number of simple potential wells (up to twelve), which can be square, parabolic, or Coulombic. You may find the eigenvalues yourself, using a "hunt and shoot" method, or allow the program to compute them automatically. You can firstly explore regular lattices, where all wells are the same and spaced at regular intervals. These will demonstrate many of the properties of regular crystals, particularly the existence of energy bands. Secondly you can change the width, depth or spacing of any of the wells, which will mimic the effect of impurities or other irregularities in a crystal. Lastly you can apply an external electric across the lattice. Facility is provided for you to incorporate your own procedures to calculate wells, lattice arrangements or external fields of their own choosing.

BOUND3D (Bound States in Three Dimensions), written by Ian Johnston, is a tool which allows you to explore energy eigenfunctions for a particle in a spherically symmetric potential well, which can be square, parabolic, Coulombic, or several other shapes of importance in molecular or nuclear applications. The first part of the program deals with finding the eigenvalues and eigenfunctions of different wells, assuming that the angular part of the wavefunctions are spherical harmonics. You may find them yourself for a given angular momentum quantum number using a "hunt and shoot" method, or else the program will compute the eigenvalues automatically, by counting the number of nodes to determine where the eigenvalues occur. The second part of the program looks at properties of eigenfunctions normalization, orthogonality, and the evaluation of many kinds of overlap integrals. Facility is provided for you to incorporate your own procedures to specify different potential wells or different overlap integrals.

IDENT (Identical Particles in Quantum Mechanics), written by Daniel Styer, shows the probability density associated with the symmetrized, antisymmetrized, or nonsymmetrized wave functions of two noninteracting particles moving in a one-dimensional infinite square well. It is particularly valuable for demonstrating the effective interaction of noninteracting identical particles due to interchange symmetry requirements.

SCATTR3D (Scattering in Three Dimensions), written by John Hiller, performs a partial-wave analysis of scattering from a spherically symmetric potential. Radial and 3-D wavefunctions are displayed, as are phase shifts, and differential and total cross sections. The analysis employs an expansion in the natural angular momentum basis for the scattering wavefunction. The radial wavefunctions are computed numerically; outside the region where the potential is important they reduce to a linear combination of Bessel functions which asymptotically differs from the free radial wavefunction by only a phase. Knowledge of these phase shifts for the dominant values of angular momentum is used to approximate the cross sections.

CYLSYM (Cylindrically Symmetric Potentials), written by John Hiller, solves the time-independent Schrödinger equation Hu=Eu in the case of a cylindrically symmetric potential for the lowest state of a chosen parity and magnetic quantum number. The method of solution is based on evolution in imaginary time, which converges to the state of the lowest energy that has the symmetry of the initial guess. The Alternating Direction Implicit method is used to solve a diffusion equation given by $HU = -\hbar \partial U / \partial t$, where H is the Hamiltonian that appears in the Schrödinger equation. At large times, U is nearly proportional to the lowest eigenfunction of H, and the expectation value $\langle H \rangle = \langle U | H | U \rangle / \langle U | U \rangle$ is an estimate for the associated eigenenergy.

SOLID STATE PHYSICS

LATCE1D (Wavefunctions for a one-dimensional Lattice), written by Ian Johnston, and included here, is described under the Quantum Mechanics heading.

SOLIDLAB (Build Your Own Solid State Devices), written by Steven Spicklemire, is a simulation of a semiconductor device. The device can be "drawn" by the user, and the characteristics of the device adjusted by the user during the simulation. The user can see how charge density, current density, and electric potential vary throughout the device during its operation.

LCAOWORK (Wavefunctions in the LCAO Approximation), written by Steven Spicklemire, is a simulation of the interaction of 2-D atoms within small atomic clusters. The atoms can be adjusted and moved around while their quantum mechanical wavefunctions are calculated in real time. The student can investigate the dependence of various properties of these atomic clusters on the properties of individual atoms, and the geometric arrangement of the atoms within the cluster.

PHONON (Phonons and Density of States), written by Graham Keeler, calculates and displays phonon dispersion curves and the density of states for a number of different 3-D cubic crystal structures. The displays of the dispersion curves show realistic curves and allow the user to study the effect of changing the interatomic forces between nearest and further neighbor atoms and, for diatomic crystal structures, changing the ratio of the atomic masses. The density of states calculation shows how the complex shapes of real densities of states are built up from simpler distributions for each mode of polarization, and enables the user to match the features of the distribution to corresponding features on the dispersion curves. In order to help with visualization of the crystal lattices involved, the program also shows 3-D projections of the different crystal structures.

SPHEAT (Calculation of Specific Heat), written by Graham Keeler, calculates and displays the temperature variation of the lattice specific heat for a number of different theoretical models, including the Einstein model and the Debye model. It also makes the calculation for a computer simulation of a realistic density of states, in which the user can vary the important parameters of the crystal, including those affecting the density of states. The program can display the results for a small region near the origin, and as a T-cubed plot to enable the user to investigate the low temperature limit of the specific heat, or in the form of the equivalent Debye temperature to enhance a study of the deviations from the Debye model. The Schottky specific heat anomaly can also be investigated.

BANDS (Energy Bands), written by Roger Rollins, calculates and displays, for easy comparison, the energy dispersion curves and corresponding wavefunctions for an electron in a 1-D symmetric $V(x) = V(-x)$ periodic potential of arbitrary shape and of strength V_0. The method used is based on an exact, non-perturbative approach so that the energy dispersion curves and band gaps can be obtained for large V_0. Wavefunctions can be displayed, and compared with one another, by clicking the mouse on the desired states on the energy dispersion curve. Changes in band structure can be followed as changes are made in the shape of the potential. The variation of the band gaps with V_0 is calculated and compared with the two opposite limits of very weak V_0 (perturbation method) and very strong V_0 (isolated atom). Even the experienced condensed matter researcher may be surprised by some of the results! Open-ended class discussions can result from the interesting physics found in these conceptually simple model calculations.

PACKET (Electron Wavepacket in a 1-D Lattice), written by Roger Rollins, shows a live animation, calculated in real time, demonstrating how an electron wavepacket in a metal or semiconducting crystal moves under the influence of external forces. The time-dependent Schrödinger equation is solved in a tight binding approximation, including the external force terms, and the motion of the wavepacket is obtained directly. The main objective of the simulation is to show that an electron wavepacket formed from states with energies near the top of an energy band is accelerated in a direction *opposite* to the direction of the external force; it has a *negative* effective mass! The simulation deals with motion in a 1-D lattice but the concepts are applicable to the full 3-D motion of an electron in a real crystal. Numerical experiments on the motion of the packet explore interesting physics questions such as: how does constant applied force affect the periodic motion of a packet? when does the usual semiclassical model fail? what happens to the dynamics of the packet when placed in a superlattice with lattice constant twice that of the original lattice?

THERMAL AND STATISTICAL PHYSICS PROGRAMS

ENGDRV, written by Lynna Spornick, is a driver program for **ENGINE, DIESEL, OTTO, and WANKEL**. These programs provide an introduction to the thermodynamics of engines.

ENGINE (Design Your Own Engine), written by Lynna Spornick, lets the user design an engine by specifying the processes (adiabatic, isobaric, isochoric [constant volume], and isothermic) in the engine's cycle, the engine type (reversible or irreversible), and the gas type (helium, argon, nitrogen, or steam). The thermodynamic properties (heat exchanged, work done, and change in internal energy) for each process and the engine's efficiency are computed.

DIESEL, OTTO, and WANKEL, written by Lynna Spornick, provide animations of each of these types of engine. Plots of the temperature versus entropy and the pressure versus volume for the cycles are shown with the engine's current thermodynamic conditions indicated.

PROBDRV, written by Lynna Spornick, is a driver program for **GALTON, POISEXP, TWOD, KAC, and STADIUM**. Subprograms GALTON, POISEXP, and TWOD provide an introduction to probability and subprograms KAC and STADIUM provide an introduction to statistics.

GALTON (A Galton Board), written by Lynna Spornick, models either a traditional Galton Board or a customized Galton Board with traps, reflecting, and/or absorbing walls. GALTON demonstrates the binominal and normal distributions, the laws of probability, and the central limit theorem.

POISEXP (Poisson Probability Distribution in Nuclear Decay), written by Lynna Spornick, uses the decay of radioactive atoms to describe the Poisson and the exponential distributions.

TWOD (2-D Random Walk), written by Lynna Spornick, models a random walk in two dimensions. A "drunk," taking equal-length steps, is required to walk either on a grid or on a plane. TWOD demonstrates the joint probability of two independent processes, the binominal distribution, and the Rayleigh distribution.

KAC (A Kac Ring), written by Lynna Spornick, uses a Kac ring to demonstrate that large mechanical systems, whose equations of motion are solvable and which obey time reversal and have a Poincaré cycle, can also be described by statistical models.

STADIUM (The Stadium Model), written by Lynna Spornick, uses a stadium model to demonstrate that there exist mechanical systems whose equations of motion are solvable but whose motion is not predictable because of the system's chaotic nature.

ISING (Ising Model in One and Two Dimensions), written by Harvey Gould, allows the user to explore the static and dynamic properties of the 1- and 2-D Ising model using four different Monte Carlo algorithms and three different ensembles. The choice of the Metropolis algorithm allows the user to study the Ising model at constant temperature and external magnetic field. The orientation of the spins is shown on the screen as well as the evolution of the total energy or magnetization. The mean energy, magnetization, heat capacity, and susceptibility are monitored as a function of the number of configurations that are sampled. Other computed quantities include the equilibrium-averaged energy and magnetization autocorrelation functions and the energy histogram. Important physical concepts that can be studied with the aid of the program include the Boltzmann probability, the qualitative behavior of systems near critical points, critical exponents, the renormalization group, and critical slowing down. Other algorithms that can be chosen by the user correspond to spin exchange dynamics (constant magnetization), constant energy (the demon algorithm), and single cluster Wolff dynamics. The latter is particularly useful for generating equilibrium configurations at the critical point.

MANYPART (Many Particle Molecular Dynamics), written by Harvey Gould, allows the user to simulate a dense gas, liquid, or solid in two dimensions using either molecular dynamics (constant energy, constant volume) or Monte Carlo (constant temperature, constant volume) methods. Both hard disks and the Lennard-Jones interaction can be chosen. The trajectories of the particles are shown as the system evolves. Physical quantities of interest that are monitored include the pressure, temperature, heat capacity, mean square displacement, distribution of the speeds and velocities, and the pair correlation function. Important physical concepts that can be studied with the aid of the program include the Maxwell-Boltzmann probability distribution, fluctuations, equation of state, correlations, and the importance of chaotic mixing.

FLUIDS (Thermodynamics of Fluids), written by Jan Tobochnik, allows the user to explore the fluid (gas and liquid) phase diagrams for the van der Waals model and water. The user chooses four phase diagrams from among the following choices: *PT, Pv, vT, uT, sT, uv,* and *sv,* where P is the pressure, T is the temperature, v is the specific volume, S is the specific entropy, and u is the specific internal energy. The program reads in the coexistence table for the van der Waals model and water, and uses it along with an empirical formula for the water free energy and the free energy derived from the van der Waals model. Given v and u, any thermodynamic quantity can be calculated. For the van der waals model thermodynamic quantities also can be calculated from the other thermodynamic state variables. The user can draw a straight line path in one phase diagram and see how this path looks in the other phase diagrams. The user also can extract all important thermodynamic data at any point in a phase diagram.

QMGAS1 (Quantum Mechanical Gas—Part 1), written by Jan Tobochnik, does the numerical calculations necessary to solve for the thermodynamic properties of quantum ideal gases, including photons in blackbody radiation, ideal bosons, phonons in the Debye theory, non-interacting fermions, and the classical limits of these systems. The user chooses the type of statistics (Bose-Einstein, Fermi-Dirac, or Maxwell-Boltzmann), the dimension of space, the form of the dispersion relation (restricted to simple powers), whether or not the particles have a non-zero chemical potential, and whether or not there is a Debye cutoff. The program then allows the user to build up a table of thermodynamic data, including the energy, specific heat, and chemical potential as a function of temperature. This data and various distribution functions and the density of states can be plotted.

QMGAS2 (Quantum Mechanical Gas—Part2), written by Jan Tobochnik, implements a Monte Carlo simulation of a finite number of quantum particles fluctuating between various states in a finite k-space (k is the wavevector). The program orders the possible energy states into an energy level diagram and then allows particles to move from one state to another according to the usual Boltzman probability distribution. Bosons are restricted so that they may not pass through each other on the energy level diagram; fermions are further restricted so that no two fermions may be in the same state; classical particles have no restrictions. In this way indistinguishability is correctly implemented for bosons and fermions. The user chooses the type of particle, the number of particles, the size and dimension of k-space, and the temperature. During the simulation the user sees a representation of the state occupancy and plots of the average energy, the instantaneous energy, and the distribution of energy among the states, also shown are results for the average energy, specific heat, and the occupancy of the ground state.

WAVES AND OPTICS PROGRAMS

DIFFRACT (Interference and Diffraction), by Robin Giles, simulates some of the fundamental wave behaviors in Fresnel and Fraunhofer Diffraction, and other Interference and Coherence effects. In particular you will be able to study diffraction phenomena associated with a point or a set of points and a slit or set of slits using the Huyghens construction. You can also use a method developed by Cornu—the Cornu Spiral—to examine diffraction from one or two slits or one or two obstacles. You can study Fresnel and Fraunhofer diffraction with a single slit or set of slits, a rectangular aperture and a circular aperture. Finally you can study Partial Coherence and fringe visibility in interference and diffraction observations. In the latter example you will be able to study the Michelson Stellar Interferometer and measure the separation distance in a double star and measure the diameter of single stars.

SPECTRUM (Applications of Interfence and Diffraction), by Robin Giles, simulates the uses and modes of operation of four important optical instruments—the Diffraction Grating, the Prism Spectrometer, the Michelson Interferometer and the Fabry-Perot Interferometer. You will look at the nature of the spectra, simulated interference patterns, and the question of resolving power.

WAVE (One-Dimensional Waves), by Wolfgang Christian, Andrew Antonelli, and Susan Fischer, uses finite difference methods to study the time evolution of the following partial differential equations: classical wave, Schrödinger, diffusion, Klein-Gordon, sine-Gordon, phi four, and double sine-Gordon. The user may vary the initial function and boundary conditions. Unique features of the program include mouse-driven drawing tools that enable the user to create sources, segments, and detectors anywhere inside the medium. Double-clicking on a segment allows the user to edit properties such as index of refraction or potential in order to model barrier problems such as thin film interference filters or the Ramsauer-Townsend effect in optics and quantum mechanics, respectively. Various types of analysis can be performed, including detector value, space-time, Fourier analysis and energy density.

CHAIN (One-Dimensional Lattice of Coupled Oscillators), by Wolfgang Christian, Andrew Antonelli, and Susan Fischer, allows the user to examine the time evolution of a 1-D lattice of coupled oscillators. These oscillators are represented on screen as a chain of masses undergoing vertical displacement. The program allows the user to examine how the application of Newtonian mechanics to these masses leads to traveling and standing waves. The relationship between the lattice spacing and other properties such as dispersion, band gaps, and cut-off frequency can be examined. Each mass can be assigned linear, quadratic, and cubic nearest neighbor interactions as well as a time-dependent external force function. Global properties such as the total energy in the lattice or the Fourier transform of the lattice can be displayed as well as the time evolution of a single mass's dynamical variables.

FOURIER (Fourier Analysis and Synthesis), written by Brian James, allows investigation of Fourier analysis and 1-D and 2-D Fourier transforms. In Fourier analysis users can choose from several predefined functions or enter their own functions either algebraically, numerically, or graphically. The build-up of a periodic function is illustrated as successive terms of the Fourier series are added in, and the effects of dispersion and attenuation on the evolution of the synthesized waveform can then be investigated. One- and two-dimensional discrete Fourier transforms can be produced for a range of standard and user-entered functions. The effects of filters on the inverse transforms are illustrated. The 2-D transforms are shown as surface and contour plots. Image processing can be illustrated by filtering the transforms of gray level images so that when the inverse transforms are displayed it can be seen that the images have been modified.

RAYTRACE (Ray Tracing and Lenses), by Brian James, lets the user explore the applications of ray tracing in geometrical optics. The fundamental principle of Fermat can be illustrated by plotting the path of a ray through two different materials between fixed points. The variation of the path of a ray through a region of changing refractive index can be used to investigate the formation of mirages. The variation of pulse delay in a fiber can be calculated as a function of its parameters and the characteristics of optical communication fibers are considered. The formation of primary and secondary rainbows due to dispersion of refractive index can be displayed. The matrix method of tracing rays through lenses can be used to investigate the images formed and show how aberrations in images arise and may be reduced.

QUICKRAY (Quick Ray Tracing), by John Philpott, can be used to demonstrate ray diagrams for a single thin lens or spherical mirror. The object and image are shown, along with the three principal rays that proceed from the object towards the observer. You can use the mouse to move the object, the position of the lens or mirror or to change the focal length of the lens or mirror. The principal rays are continuously redrawn while any of these adjustments are made. The simulation handles converging and diverging lenses and concave and convex mirrors. Thus students can quickly get an intuitive feel for real and virtual image formation under a variety of circumstances.

Acknowledgments

The CUPS Project was funded by the National Science Foundation (under grant number PHY-9014548), and it has received support from the IBM Corporation, the Apple Corporation, and George Mason University.

2

Scalar and Vector Fields

Robert Ehrlich and Jaroslaw Tuszynski

2.1 Introduction

To obtain a quick introduction to this simulation see the FIELDS "Walk Through" in Appendix A. A scalar (or vector) field is a scalar (or vector) function defined over some domain in space. Scalar and vector fields may be used to represent a great many physical quantities in various areas of physics, including electricity and magnetism. In this chapter we are concerned with the mathematics of scalar and vector fields, without attaching any particular significance to the physical quantities they represent. We shall use the symbols $\vec{A}$ (or F) to refer to an arbitrary vector (or scalar) field. Our discussion here shall be very brief, and shall give in outline form what you can find in the introductory chapter of many standard texts on electricity and magnetism.[1-6]

2.2 Differentiation of Fields

2.2.1 Scalar Fields

For a scalar field F, we can define a special type of differentiation, known as the *gradient*, abbreviated grad, and defined in Cartesian coordinates by

$$grad\, F = \hat{i}\frac{\partial F}{\partial x} + \hat{j}\frac{\partial F}{\partial y} + \hat{k}\frac{\partial F}{\partial z},\qquad(2.1)$$

where $\hat{i}$, $\hat{j}$, and $\hat{k}$ are the unit vectors directed along the x-, y-, and z-axes. It is very convenient to define the *grad operator* $\vec{\nabla}$:

$$\vec{\nabla} = \hat{i}\frac{\partial}{\partial x} + \hat{j}\frac{\partial}{\partial y} + \hat{k}\frac{\partial}{\partial z},\qquad(2.2)$$

so that we may write *grad F* = $\vec{\nabla}F$. Note that by taking the gradient of a scalar field *F*, we produce a new field $\vec{A} = \vec{\nabla}F$ that is a vector field. Geometrically, if we represent a two-dimensional (2-D) scalar field by a contour map, where the contours show evenly spaced constant values of the function *F*, the gradient is inversely proportional to the spacing between the contours, and it points along the path of steepest ascent. There are various ways to display the vector field $\vec{\nabla}F$. Two of the more common methods are to use a collection of vectors shown at points on a grid, and a *field line* representation, where the density of field lines at each point shows the magnitude of the field there, and a tangent to a field line shows its direction.

In addition to the gradient, another useful quantity is the *Laplacian* of an arbitrary scalar field *F*, which is defined, in a formal sense, as the dot product of $\vec{\nabla}$ with $\vec{\nabla}F$. Thus:

$$\vec{\nabla} \cdot \vec{\nabla}F = \nabla^2 F = \frac{\partial^2 F}{\partial x^2} + \frac{\partial^2 F}{\partial y^2} + \frac{\partial^2 F}{\partial z^2}, \tag{2.3}$$

2.2.2　Vector Fields

The grad operator can also be used to define two other useful quantities, namely, the *divergence* and *curl* of a vector:

$$div\ \vec{A} = \vec{\nabla} \cdot \vec{A} = \frac{\partial A_x}{\partial x} + \frac{\partial A_y}{\partial y} + \frac{\partial A_z}{\partial z}, \tag{2.4}$$

and

$$curl\ \vec{A} = \vec{\nabla} \times \vec{A} = \hat{i}\left(\frac{\partial A_z}{\partial y} - \frac{\partial A_y}{\partial z}\right) + \hat{j}\left(\frac{\partial A_x}{\partial z} - \frac{\partial A_z}{\partial x}\right) + \hat{k}\left(\frac{\partial A_y}{\partial x} - \frac{\partial A_x}{\partial y}\right). \tag{2.5}$$

The divergence of a vector field has a simple geometrical interpretation that accounts for its name: it gives the density of "sources" or "sinks" of the field at each point. For example, a vector field that is divergence-free $\vec{\nabla} \cdot \vec{A} = 0$ everywhere must have no places where field lines originate or terminate; i.e., the field lines must form closed loops. Note that by taking the divergence of a vector field, we produce a scalar field $\vec{\nabla} \cdot \vec{A}$, whereas by taking the curl, we produce another vector field, $\vec{\nabla} \times \vec{A}$. Although we have seen in the previous section the combined operator $\vec{\nabla} \cdot \vec{\nabla}$, there is no operator corresponding to the combination $\vec{\nabla} \times \vec{\nabla}$ for scalar fields, because as long as the order of partial differentiation does not matter, $\vec{\nabla} \times \vec{\nabla}F$ must vanish. We may call such fields "curl-free" or "irrotational."

Although we have here defined the gradient, divergence, curl, and Laplacian operators in Cartesian coordinates, they may also be defined in other coordinate systems. Formulas defining these operators in various coordinate systems can be found in any of the standard textbooks. Two formulas which shall prove useful in working several exercises include those for the Laplacian in cylindrical and spherical coordinates. In cylindrical coordinates (r,θ,z),

$$\nabla^2 F = \frac{1}{r}\frac{\partial}{\partial r}\left(r\frac{\partial F}{\partial r}\right) + \frac{1}{r^2}\frac{\partial^2 F}{\partial \theta^2} + \frac{\partial^2 F}{\partial z^2}, \tag{2.6}$$

and in spherical coordinates ρ, θ, ϕ, we have

$$\nabla^2 F = \frac{1}{\rho^2}\frac{\partial}{\partial\rho}\left(\rho^2\frac{\partial F}{\partial r}\right) + \frac{1}{\rho^2 \sin\theta}\frac{\partial}{\partial\theta}\left(\sin\theta\frac{\partial F}{\partial\theta}\right) + \frac{1}{\rho^2 \sin^2\theta}\frac{\partial^2 F}{\partial\phi^2}. \qquad (2.7)$$

2.3   Integration of Fields

Three types of integrals are particularly important, namely, line and surface integrals of vectors, and volume integrals of scalars.

2.3.1   Line Integrals

Let $\vec{A}$ be an arbitrary vector field, and let $\vec{dl}$ be an infinitesimal displacement along some particular path connecting point a and point b. Using the definition of the dot product of two vectors, we may write a line (or path) integral as

$$\int_a^b \vec{A}\cdot\vec{dl} = \int_a^b (A_x\,dx + A_y\,dy + A_z\,dz), \qquad (2.8)$$

Depending on the field $\vec{A}$, it may happen that the result of the integral is independent of the particular path—in which case the field $\vec{A}$ is said to be a *conservative* field. Conservative fields also have the property that if the path is closed, i.e., if a and b are the same point, then the integral must be zero. You probably are familiar with one important physical application of line integrals, namely, the work done by a force $\vec{F}$ in moving an object from point a to point b:

$$W = \int_a^b \vec{F}\cdot\vec{dl}. \qquad (2.9)$$

Forces such as gravity that result in a path-independent result are known as conservative forces. In electricity and magnetism we shall encounter both conservative and non-conservative forces.

2.3.2   Surface Integrals

Let $\vec{A}$ be an arbitrary vector field, and let $\vec{dS}$ be an infinitesimal element of some surface over which we want to integrate the field. For example, if the surface were to lie entirely in the x-y plane, then in magnitude $dS = dx\,dy$. We take the direction of $\vec{dS}$ to be perpendicular to the surface at each point on it. Using the definition of the dot product of two vectors, we may define the integral of a vector $\vec{A}$ over a surface:

$$\int_S \vec{A}\cdot\vec{dS} = \int_S A\,dS\,\cos\theta \qquad (2.10)$$

where θ is the angle between $\vec{A}$ and $\vec{dS}$. For an integral over a closed surface, the direction of $\vec{dS}$ is that of the *outward* normal. In the simple case where $\vec{A}$ has a constant magnitude, and makes a constant angle with the surface normal, the integral becomes $AS \cos\theta$. Thus, the surface integral of the position vector $\vec{r}$ over the closed surface of a sphere of radius 2 centered on the origin would be 32π. On the other hand, the integral of a uniform vector field over the same sphere would have to be zero. Do you see how the result follows from left-right symmetry without having to carry out the integral?

2.3.3   Volume Integrals

The volume integral of an arbitrary scalar field over some closed region V may be found in Cartesian coordinates using a triple integral:

$$\int_V F\,dv = \iiint F\,dx\,dy\,dz. \tag{2.11}$$

Depending on the nature of the scalar field, it may be more convenient to do the integral in another coordinate system, such as cylindrical or spherical. For example, in spherical coordinates:

$$\int_V F\,dv = \iiint F r^2\,dr\,d\phi\,d(\cos\theta), \tag{2.12}$$

which for a spherically symmetric field (independent of ϕ and θ) would reduce to $4\pi \int F r^2\,dr$.

2.3.4   Stokes' and Gauss' Theorems

There are many relationships involving integrals of vector fields. Among the more useful ones are *Stokes'* and *Gauss'* theorems, which are stated below without proof. Stokes' theorem states that the line integral of a vector field $\vec{A}$ around a closed path equals the surface integral of the curl of $\vec{A}$ over a surface bounded by that path, or in symbols:

$$\oint \vec{A} \cdot d\vec{l} = \int (\vec{\nabla} \times \vec{A}) \cdot d\vec{S}. \tag{2.13}$$

The sign convention in Stoke's theorem follows the right-hand rule: when the fingers of your right hand show the sense in which the line integral is performed, your thumb points in the direction of the surface normal. Gauss' theorem states that the surface integral of a vector field $\vec{A}$ over some closed surface equals the volume integral of the divergence of $\vec{A}$ inside that closed surface, or in symbols:

$$\oint \vec{A} \cdot d\vec{S} = \int (\vec{\nabla} \cdot \vec{A})dv. \tag{2.14}$$

To take one example of Stokes' theorem, suppose a given vector field had a curl of $5\hat{k}$ everywhere inside a unit circle in the x-y plane. According to Stokes' theorem, its line integral around the circle would be 5π. Note that Stokes' theorem implies that a curl-free field must be a conservative field, since its line integral around a closed path must be zero. Another implication of Stokes' theorem is that a vector field whose field lines form closed loops cannot have zero curl everywhere, because if we integrate the field along a path coincident with a closed field line the integral cannot be zero, so the integral of the curl must also be non-zero. But here again, the converse need not follow: a field with non-zero curl need not form closed loops.

2.4 Running FIELDS

2.4.1 Introduction

The FIELDS program allows you to enter any scalar or vector field that can be specified in terms of an expression involving functions recognizable to a function parser, such as exp, sin, cos, log, sqrt, abs. See the preface of this book for a complete list of functions, and how they need to be entered into the program. For vector fields you need to specify separate expressions for A_x, A_y, and A_z, while for a scalar field only one expression need be specified. Once you have specified the field, and the domain in x, y, and z over which the expressions are to be evaluated, you then need to select four plots to be displayed.

2.4.2 Computation and Display Options

In order to indicate what you want the program to display, you need to select **Fields | Choose Graphs**. For a scalar field F, or a vector field $\vec{A}$, you may choose any four plots from the tabulated quantities or you could select one plot to be a "mouse probe" or a path integral window (Table 2.1).

Table 2.1: Selectable Plots

Scalar Field	Vector Field
F	$\vec{A}$
$\vec{\nabla}F$	$|\vec{A}|$
$|\vec{\nabla}F|$	$A_{x,y,z}$
$(\vec{\nabla}F)_x$	$\vec{\nabla} \cdot A$
$(\vec{\nabla}F)_y$	$\vec{\nabla} \times A$
$(\vec{\nabla}F)_z$	$|\vec{\nabla} \times A|$
$\vec{\nabla}^2 F$	$(\vec{\nabla} \times A)_{x,y,z}$

The partial derivatives used in the expressions for the gradient, divergence, curl, and Laplacian are all calculated numerically (not analytically). To do this, the field is first evaluated numerically at each point on finite two-dimensional grids whose size may be specified. (The default grid size in the program is 30 by 30.) For derivatives involving dimensions perpendicular to the plane of the grid, the field is evaluated at a total of five grid planes—including two on either side of the plane in question. The partial derivatives are found at each grid point (x, y, z) using the following numerical approximations:[7]

$$\frac{\partial F}{\partial x} = \frac{1}{24\Delta x}[2F_{-2} - 16F_{-1} + 16F_1 - 2F_2] + O(\Delta x^5) \qquad (2.15)$$

$$\frac{\partial^2 F}{\partial x^2} = \frac{1}{12\Delta x^2} [-F_{-2} + 16F_{-1} - 30F_0 + 16F_1 - F_2] + O(\Delta x^6), \quad (2.16)$$

where $F_k = F(x + k\Delta x, y, z)$. Completely analogous expressions are used for partial derivatives with repect to y and z. These formulas will, of course, yield erroneous results if they are applied in the vicinity of a singularity.

Finally, you need to choose in which plane you want the field displayed, using a separate menu item: **Fields | Choose Plane**. You may select any plane at constant x, y, or z, i.e., any planes parallel to the x-y, x-z, or y-z coordinate planes. There is no need for you to specify the domain of the display, since this was specified at the time you entered expressions for the field.

The default method of display for scalar fields is the familiar contour map, where color is used to code the value of each contour. For example, a uniform scalar field would produce no contours, while a field that varied linearly with the x-coordinate would produce equally spaced parallel contours perpendicular to the x-direction. You can also switch from a contour map display to a three-dimensional (3-D) surface display. By clicking the mouse on the small square in the upper left corner of the graph you can change back and forth between the 3-D and contour displays. For a 3-D plot of a scalar field, the field value at each point in the plane is indicated by the height of a 3-D surface, which can be rotated using sliders and viewed from various perspectives. (The sliders adjust the first and second Euler angles, i.e., the azimuth around the z-axis, and the azimuth around the new x-axis.) Note, that the maximum height of the 3-D surface is automatically adjusted to give a convenient display.

The method used to display vector fields in a 2-D plot is to use small arrows attached to the grid sites in the plane in which the field is calculated. The direction of the arrows shows the field components in the plane being displayed, their length is scaled according to the magnitude of those components in the plane, and the color of the arrows is used to code the field component perpendicular to the plane shown. A field that only has components perpendicular to the plane of the plot will not produce any display. For a 3-D plot of a vector field, the little vectors are color-coded according to their length. As with the scalar field, you can switch between the two display options by clicking on the small square at the corner of a plot.

The "mouse probe" display window gives you numerical values for a range of useful quantities at the current location of the mouse in whichever 2-D window it is located. These values are found, using a standard six-point interpolation procedure, from the values of the six grid points nearest to the current position. Of course, any interpolation procedure may give imprecise results—particularly for rapidly varying functions. In some cases, therefore, you may wish to decrease the grid spacing (increase the grid size), using the menu option **Fields | Resize Matrix**—at the price of slowing down the calculation.

You may also choose to make one of your four plots a "path integral window." In that case, if you perform a path integral in one of the other plots (by clicking the mouse at a series of points denoting the ends of line segments), the results of the integral will be displayed in the path integral window. The program does two types of path integrals: open and closed. The open path consists of one or more line segments, and the closed path consists of an arbitrary polygon—both being defined using the mouse. If you want to perform a line integral along

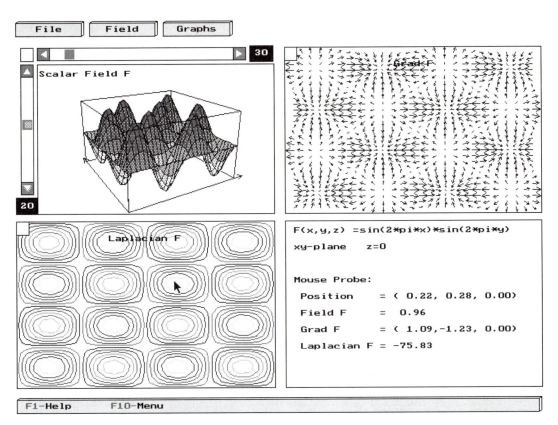

Figure 2.1: The scalar field $F = \sin 2\pi x \sin 2\pi y$ displayed as a 3-D surface, its gradient displayed using little arrows, and its Laplacian displayed as a contour plot. The lower right quadrant shows the mouse probe window, with the numerical values of the indicated quantities corresponding to the current position of the mouse.

a particular path defined by $y = f(x)$ you would need to approximate that path by a series of straight line segments connecting points having specific coordinates. (You would actually need to evaluate those coordinates numerically ahead of time so as to be able to move the mouse in the appropriate manner, as explained in one of the exercises.) Two additional methods for carrying out this procedure for arbitrary functions, $y = f(x)$, that do not require such a numerical evaluation are also suggested in the exercises.

2.5 Exercises

2.5.1 Scalar Fields in Cartesian Coordinates

2.1 Consider the scalar function $F = e^{-x^2/2 - y^2}$. Analytically, find where the magnitude of the gradient is a maximum, and where it is zero. Also, find analytically where the Laplacian is a maximum. Run the program, and enter the expression **exp(-0.5*x^2-y^2)**. Use the mouse probe to measure the points where $|\vec{\nabla}F|$ has a maximum and zero, and also see where $\vec{\nabla}^2 F$ is a maximum. How close did you come to your analytical results?

2.2 Consider the scalar function $F = \sin 2\pi x. \sin 2\pi y$. Prove analytically that $\vec{\nabla}^2 F = -F$. Demonstrate that this is true using the program by displaying both $\vec{\nabla}^2 F$ and F. Using the mouse probe, measure the values of F and $\vec{\nabla}^2 F$ at five randomly chosen points, and see how closely the two functions agree, apart from a minus sign. Also, find analytically the lines along which $\vec{\nabla}F$ has no x-component, and use the program to confirm your results.

2.3 Consider the scalar function $F = \sin 2\pi x. \sin 2\pi y$ to which we add some random noise of amplitude z. This would be entered into the program as: **sin(2*pi*x)*sin(2*pi*y)+0.1*z*rand.** Examine the plots that the program produces in x-y planes corresponding to z-values from 0 to 0.2 (different noise levels). Explain why the three plots corresponding to F, $|\vec{\nabla}F|$, and $\vec{\nabla}^2 F$ are affected to different degrees by the addition of noise. For what z-values do you just begin to notice the effects of noise in each of the three plots? Is the effect of noise on $|\vec{\nabla}F|$ greatest near maxima or zeroes of F?

2.4 In the previous exercise random noise was added to the amplitude of a periodic function F. Suppose instead of this "amplitude noise," we were to add some "phase noise." In this case, the function to be entered into the program would be **sin(2*pi*(x+0.1*z*rand))*sin(2*pi*(y+0.1*z*rand)).** Examine the plots that the program produces in x-y planes corresponding to z-values from 0 to 0.2. Is the effect of phase noise on $|\vec{\nabla}F|$ greatest near maxima or zeroes of F? What other differences, if any, do you find in the effects of the two types of noise? Explain why these specific differences occur.

2.5 Consider the scalar function $F = \sin 200\pi x. \sin 200\pi y$. Use the program to display the function inside the region of the x-y plane defined by the default limits. What would you have expected the plot to look like? Account for any differences between the plot and your expectation. Hint: Display the function using twice as fine a grid. The effect you are observing here is known as "aliasing."

2.6 Consider the scalar function $F = e^{-(x-0.5)^2-y^2/2}$. Find analytically where the gradient has its greatest x-component. Confirm your result by running the program, and finding the coordinates where the mouse probe shows the maximum value.

2.7 Consider the scalar function $F = x^3 + y^3 + z^3$. Calculate analytically $\vec{\nabla}^2 F$, and show that it has a constant value on lines of 135-degree slope. Use the program to confirm your results. Are the lines of constant $\vec{\nabla}^2 F$ evenly spaced? Explain.

2.8 Consider the scalar field $F = e^{-5(x-0.5)^2-y^2} + e^{-5(x+0.5)^2-y^2}$. Without doing the calculation, explain why it is plausible that F should have two maxima very close to $(0.5, 0.0)$ and $(-0.5, 0.0)$. Explain why it is plausible that
the Laplacian should also have maxima or minima at those same two
points. Use the mouse probe to test these assertions.

2.9 Consider a collection of tiny magnets arranged in a square array. Let us
assume that each magnet is able to align itself in only one of two directions, which we may call "up" or "down." This can serve as a crude model
of a collection of elementary particles such as electrons in a solid. If the
particles or tiny magnets obey so-called "Fermi-Dirac" statistics, the fraction of magnets with their spins pointing in one direction can be expressed
as $f(z) = 1/(e^{-z} + 1)$, where z is the inverse of the temperature of the system divided by some critical temperature. Consider the function $h(f(z) -
rand)$. Recall that $h(x)$ is defined to be 1 for x greater than zero, and zero
otherwise. The given function therefore has the property that it will have
the value 1.0 ("spin up") a fraction $f(z)$ of the time, and the value 0.0 ("spin
down"), the rest of the time. Using the program, enter the vector field
$A_x = 0, A_y = h(1/(e^{-z} + 1) - rand), A_z = 0$, for a range of z-values from
0 to 2. Observe the field in x-y planes corresponding to a series of z-planes
from $z = 0$ to $z = 2$. In each plane estimate the fraction of random spins
with spin up and spin down. How does the fraction of atoms with spins up
vary with z? What is the fraction for $z = 0$ (infinite temperature)? For
what z-value are half the spins pointing up?

2.5.2 Scalar Fields in Other Coordinates

2.10 Consider the scalar function $F = 1/(1 + e^{-(r-0.5)/\Delta r})$. Run the program
and qualitatively describe how the function changes as Δr decreases. Try
values for $\Delta r = 0.4, 0.1.,$ and 0.01. (Note that as Δr approaches zero, the
function approaches the unit step function $h[0.5 - r]$.) Show analytically
that the divergence has its maximum magnitude at $r = 0.5$. Use the mouse
probe to see where the plot of $|\vec{\nabla} F|$ shows a maximum.

2.11 Use the Laplacian in spherical coordinates to show analytically that for
$F = r^n, \nabla^2 F = n(n + 1)r^{n-2}$. Check the correctness of this result by running the program using $F = r^3$, and examining the plot of the Laplacian.
What aspect of the plot shows that the result is correct? Repeat the exercise
for cylindrical coordinates.

2.12 Consider a scalar field in cylindrical coordinates: $F = r^2\phi^2$. Use
the Laplacian in cylindrical coordinates to show analytically that $\vec{\nabla}^2 F =
2 + 4\phi^2$. Run the program to display the Laplacian, and check the ϕ de-
pendence by recording $\vec{\nabla}^2 F$ at four values of ϕ using the mouse probe.
(Even though the mouse probe does not show the value of ϕ, you can easily
tell when you are at certain fixed angles based on the ratio of the x- and
y- coordinates.)

2.13 Consider the scalar function $F = h(0.8 - r)$. What should $|\vec{\nabla}F|$ and $\nabla^2 F$ look like? Discuss the reason the plots differ from your expectations. (Hint: Try increasing the number of grid points used in the plots.) What is the width of the region over which $\vec{\nabla}^2 F$ is non-zero in units of the grid spacing? Ideally, what should it be? Explain.

2.14 Show analytically that the Laplacian of the scalar function $F = -1/r$ is a "delta function," i.e., a spike of infinite height and zero width centered on $r = 0$. Run the program using several different grid sizes, and evaluate the width of the spike at half its maximum height for each grid spacing. Is the width proportional to the grid spacing?

2.15 Using cylindrical coordinates, show analytically that the Laplacian of the scalar field $F = r^2$ is a constant. Use the program to verify this.

2.5.3   Vector Fields

2.16 Find a vector field whose divergence is 3.0, and whose curl is zero. Is your answer unique? If not, write several other examples. Use the program to check your answer in each case.

2.17 Consider the vector field $A_x = A_y = A_z = r$. Use the program to show that the divergence and curl are nearly independent of r, and confirm this result analytically. Based on the plots, along what directions is the divergence a maximum, and a minimum?

2.18 Consider an arbitrary vector field (A_x, A_y, A_z). In a coordinate system x', y' rotated in the x-y plane, the components may be written in terms of the old components as $A'_x = A_x \cos \theta + A_y \sin \theta$, and $A'_y = -A_x \sin \theta + A_y \cos \theta$. Choose a vector field that lacks cylindrical symmetry, such as $A_x = r$, $A_y = x$. Write down the components of the vector field after a rotation by 45 degrees. Enter the rotated vector field and the original vector field in the program and compare the plots. If they are not exactly what you expected, account for any differences.

2.19 Consider the vector field $A_x = yz$, $A_y = -xz$, $A_z = 0$. Show analytically that its curl is $x\hat{i} + y\hat{j} - 2z\hat{k}$. Use the program to verify the result by displaying the curl in three mutually perpendicular planes, and using the mouse probe to find its value at a number of points.

2.20 Show analytically that a vector field defined by $A_x = r \cos(\theta + \pi/2)$, $A_y = r \sin(\theta + \pi/2)$, $A_z = 0$, will be everywhere tangent to concentric circles about the origin. Run the program to verify your result. Use the program to do a line integral around various closed paths to see if the field is conservative (zero integral around a closed loop). Does the value of the integral depend on what path you used?

2.21 Any curl-free vector field can be expressed as the gradient of a scalar field: $\vec{A} = \vec{\nabla}F$. Consider a vector field of the form: $A_r = r$ (for $r < 1$), and

$A_r = 1/r^2$ (for $r > 1$). Find a function F such that $\vec{A} = -\vec{\nabla}F$. Use the program to check your guess by first displaying $\vec{A}$ and then displaying $-\vec{\nabla}F$, and comparing the two at various values of r. Note that to enter a function that has the r-dependence $A_1(r)$ for $r < 1$ and $A_2(r)$ for $r > 1$, use the combination $A_1(r)h(1 - r) + A_2(r)h(r - 1)$. (Note that all fields must be entered in the program in terms of their x-, y-, and z-components.)

2.22 Show that any vector field that satisfies $\vec{A} = \vec{\nabla}F$ must have zero curl. Make up three scalar functions, and for each calculate $\vec{A} = \vec{\nabla}F$ analytically in Cartesian coordinates. Enter these fields in the program to verify that they are curl-free.

2.5.4   Line Integrals

2.23 Given a vector field $A_x = x^2$, $A_y = 2y^2 + xy$, $A_z = 0$, analytically evaluate its line integral from $(1, 2)$ to $(3, 3)$, and use the program to check your answer.

2.24 Given a vector field $A_x = \sin y$, $A_y = x(1 + \cos y)$, $A_z = 0$, analytically evaluate its curl. Also, analytically evaluate the line integral of $\vec{A}$ from $(0, 0)$ to $(1, 1)$, and use the program to check both results.

2.25 Consider a vector field defined by $A_x = -y/r$, $A_y = x/r$, $A_z = 0$. Analytically evaluate its line integral over a semicircular path in the upper half of the x-y plane from $(1, 0)$ to $(-1, 0)$. Do the same for the semicircle lying in the lower half of the x-y plane. Is the line integral path-independent? Explain.

In order to use the program to check these results we need to have a simple way to draw the integration path on the screen. One way to do this is to calculate the x, y-coordinates of a number of points on the unit semicircle. For example, taking seven points 30 degrees apart, we have $(1, 0)$, $(.866, .5)$, $(.5, .866)$, $(0, 1)$, $(-.5, .866)$, $(-.866, .5)$, and $(-1, 0)$. You could then connect these seven points by moving the mouse, thereby approximating a path along the semicircle, so as to do the path integral. How do the results from this method agree with your analytical results?

2.26 In previous exercises we used a rather crude method for approximating line integrals along a continuous path using straight line segments. A completely different approach is to apply a transformation to the original field such that the path to be drawn becomes a simple straight line that can easily be drawn using the mouse. For example, suppose we wish to integrate the vector field $A_x = x^2y$, $A_y = xy^2$, $A_z = 0$ along the semicircle in the upper half plane that connects $(-1, 0)$ to $(1, 0)$. The equation of this semicircle is $y = \sqrt{1 - x^2}$. Thus, if we transform the original field by replacing y by $y - \sqrt{1 - x^2}$, then the line integral of the transformed field along the straight line joining $(-1, 0)$ and $(1, 0)$ is identical to that of the original field along the semicircle. Essentially, the field points originally lying on the semicircle have all had their y-coordinates dropped by the appropriate amounts to place them on the straight line.

Follow the above procedure for transforming a field to make the integration path a straight line, and use the program to find the line integral of the above field along a curve joining $(-1, 0)$ and $(1, 0)$, where the curve is (a) the semicircle $y = \sqrt{1 - x^2}$, (b) the parabola $y = 1 - x^2$, and (c) the cosine function $y = \cos \pi x/2$. Are the results path independent for this field?

2.27 Using the method described in the previous exercise, find a line integral around a closed path, say, the unit circle. In order to do this, you will need to break the circle up into two semicircles lying in the upper and lower half-planes, and evaluate the two line integrals separately. Come up with two fields, one of which you suspect to be conservative, and the other non-conservative, and use the program to evaluate their line integrals around the unit circle following the procedure outlined above.

2.5.5   Stokes' Theorem

2.28 Consider the vector field: $A_x = -y$, $A_y = x$, $A_z = 0$. Analytically, find its line integral around a square of unit side centered on the origin, and run the program to verify your result. Show that analytically the z-component of the curl of this field is $2\hat{k}$, and see how closely the results of the program conform to Stokes' theorem. (Hint: What is the surface integral of a constant inside the unit square?)

2.29 Can you construct a cylindrically symmetric field whose curl is everywhere given by $r^3\hat{k}$? (Hint: use Stokes' theorem.) Use the program to see if the field you came up with does in fact have the proper curl. Exactly how did you verify this?

Bibliography

1. Lorrain, P. *Electromagnetic Fields and Waves.* 3rd ed. San Francisco: W. H. Freeman & Company, 1988.

2. Reitz, H. R., Milford, F. J., and Christy, R. W. *Foundations of Electromagnetic Theory.* 3rd ed. Reading, MA: Addison-Wesley Publishing Company, 1979.

3. Griffiths, D. *Introduction to Electrodynamics.* 2nd ed. Englewood Cliffs, NJ: Prentice Hall Publishing Company, 1989.

4. Marshall, S. V. and Skitek, G. G. *Electromagnetic Concepts and Applications.* 3rd ed. Englewood Cliffs, NJ: Prentice Hall Publishing Company, 1990.

5. Visscher, P. B. *Fields and Electrodynamics.* New York: John Wiley & Sons, 1988.

6. Wangsness, R. K. *Electromagnetic Fields.* 2nd ed. New York: John Wiley & Sons, 1986.

7. Abramowitz, M. and Stegun, I., eds. *Handbook of Mathematical Functions.* Washington, DC: U.S. Department of Commerce, 1970, p. 914.

3

Poisson's Equation and Gauss' Law

Robert Ehrlich and Jaroslaw Tuszynski

3.1 Introduction

In free space we can solve for the electric field in the presence of a static charge distribution using Gauss' law:

$$\vec{\nabla}.\vec{E} = \frac{\rho}{\epsilon_0}. \tag{3.1}$$

Writing $\vec{E} = -\vec{\nabla}V$, we obtain Poisson's equation:[1-5]

$$\nabla^2 V = -\frac{\rho}{\epsilon_0}. \tag{3.2}$$

Two simulations, GAUSS and POISSON, have been written to solve these equations.

A particularly simple situation involves spherically symmetric charge distributions, where $\rho(x, y, z)$ can be written as $\rho(r)$. In this case, Gauss' law can be immediately integrated to yield

$$E(r) = \frac{q_{in}}{4\pi\epsilon_0 r^2}, \tag{3.3}$$

where q_{in} is the net electric charge inside a sphere of radius r. $E(r)$ can be similarly found for other situations having a high degree of symmetry (cylindrical and planar). Once $E(r)$ has been found by integration, we can easily find $V(r)$. In fact, given any one of the three functions $\rho(r)$, $E(r)$, and $V(r)$, the other two can be immediately found by integration or differentiation. This is the procedure carried out in the GAUSS program, which is restricted to highly symmetric problems involving continuous distributions of electric charge in free space.

A much more powerful technique is to solve Poisson's equation directly. For simplicity, we only treat two-dimensional (2-D) problems in which the potential may be written as $V(x, y)$. Physically, this restricts us to objects that are long ("infinite") in extent along the z-axis, such as line charges, and cylinders of various cross section. We first discretize $V(x, y)$ by defining the function on a grid of points, and for simplicity take the grid spacing to be one unit in each dimension. In this case, we have for the lowest order expressions for the second derivatives,

$$\frac{\partial^2 V}{\partial x^2} = V(i + 1, j) - 2V(i, j) + V(i - 1, j) \tag{3.4}$$

and

$$\frac{\partial^2 V}{\partial y^2} = V(i, j + 1) - 2V(i, j) + V(i, j - 1). \tag{3.5}$$

Thus, by adding the two previous expressions, we see that

$$\nabla^2 V = 4(\overline{V} - V) \tag{3.6}$$

where

$$\overline{V}(i, j) = \frac{1}{4}(V(i + 1, j) + V(i - 1, j) + V(i, j + 1) + V(i, j - 1)), \tag{3.7}$$

is the average of the potential at the four grid points nearest to (i, j). Thus, in discretized form, we can write Poisson's equation as: $4(\overline{V} - V) = -\frac{\rho}{\epsilon_0}$, where all terms are evaluated at grid points only. For simplicity, we shall use units for electric charge density such that $\epsilon_0 = 1$, which gives for the last result—

$$V(i, j)_{new} = \overline{V}(i, j) + \frac{1}{4}\rho(i, j). \tag{3.8}$$

The process of solving Eq 3.8 iteratively for $V(i, j)$ is known as the "relaxation method."[6] To begin the iteration process, the potential $V(i, j)$ at all grid points is set to zero, except at those grid points having some specified potential that is held constant from iteration to iteration. The potential at the rectangular boundaries of the grid is also held fixed during the iteration process. One choice for the boundary condition is to set V to zero at the four edges, which is equivalent to defining the system in a rectangular grounded box. Another choice in the program is to set the left edge of the box to one constant potential, the right edge to another, and the top and bottom edges to a linearly varying potential that matches those of the left and right edges—the "parallel plate capacitor" option. If any conductors are drawn on the grid, all grid points which lie on their surface must also have their potential held fixed during the iteration process. Since we wish to treat cases in which conductors have various shapes, some error of "discretization" is made when a conductor's shape is translated to the grid.

In order to speed up the iteration process, normally we use an "over-relaxation" method, in place of the standard algorithm. The idea is to push the solution at each iteration further in the direction it would move by the relation method itself. The equation describing the algorithm is

$$V(i, j)_{new} = V(i, j) + \omega(\overline{V}(i, j) - V(i, j) + \frac{1}{4}\rho(i, j)). \tag{3.9}$$

The parameter ω defines the extent of "over-relaxation," where $\omega = 1$ means no over-relaxation. Although the over-relaxation algorithm may accelerate the convergence process, it can lead to instabilities if the ω parameter is too large — values larger than 2.0 don't converge. In applying the algorithm the updated and original potentials are stored in the *same* matrix, so that the potential at each site (i, j) use values from neighboring sites that have already been updated during that iteration. By using the same matrix, information can "flow" faster through the grid, and the convergence process is speeded up.

Poisson's equation has a wide variety of physical applications besides electrostatics. For example, it can be shown that the following three problems can all be solved by Poisson's equation, and they have the same solution:

- Two conducting parallel plates at potentials $+100$ V and -100 V are at the center of a grounded box, and parallel to its sides. Find the potential everywhere in the box.

- Two parallel plates maintained at temperatures $+100^0$ C and -100^0 C are at the center of a box, and parallel to its walls. If the walls are maintained at 0^0 C, find the temperature at all points inside the box.

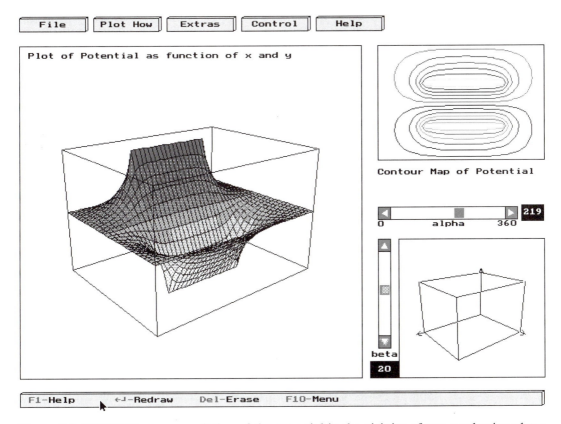

Figure 3.1: POISSON program solution of the potential in the vicinity of two conducting plates at $+100V$ and -100 at the center of a grounded box. The display on the left uses a 3-D representation, and the display on the upper right uses a contour map representation.

● An elastic rubber membrane is attached to a square frame. One plate, which is parallel to an edge of the frame, pushes from below the membrane, raising it to a height of 100 cm. A second parallel plate pushes from above the membrane, depressing it by 100 cm. Find the height of the membrane relative to the frame everywhere.

Note how the 3-D displays generated in the POISSON program to display the potential give an immediate intuitive connection to problems of this last type.

3.2   Running GAUSS

Of the two programs written for this chapter, GAUSS and POISSON, the former is by far the simpler, and so it will be described first. GAUSS is used to solve problems involving continuous charge distributions that possess a high degree of symmetry. Specifically, it can be used for any charge distribution that has spherical, cylindrical or planar symmetry. By planar symmetry, we mean that the charge distribution can be expressed in terms of a single rectangular coordinate, x. You first should specify which of the three coordinate systems is applicable to the problem—using the **Symmetry** menu item. Then you have the option of specifying a charge density function, $\rho(r)$, in terms of some known function.

For example, $h(5 - r)$, where h is the unit step function, would specify a solid sphere (or cylinder) of charge of radius 5 units having constant density, or a uniform slab of charge 5 units thick. Obviously, distributions with varying charge density could be specified in terms of other functions, such as exp, sin, etc. The input screen help shows the allowed types of functions, and the notation to be used. After you specify the charge density function $\rho(r)$, and its range, the program computes, by numerical integration of Gauss' law, the electric field $E(r)$, and the associated potential $V(r)$. If the function being input has a singularity, be aware that the computed quantities in the immediate vicinity of the singularity will not be correct.

The program displays the three functions $\rho(r)$, $E(r)$, and $V(r)$ on a common plot. The plots are automatically scaled in both dimensions to take up the full plotting area. The program also displays a dot-density plot at the bottom to convey the charge distribution pictorially. Different colors are used to represent regions of positive and negative charge density. (You need to hit any key or click the mouse to stop the dots from accumulating.) You also have a choice of displaying the results as three-dimensional figures of revolution by clicking the F2 hot key. The 3-D surfaces may be viewed from any perspective by adjusting two sliders.

In addition to allowing you to input a particular $\rho(r)$, the program also allows you to enter functional forms for either $E(r)$ or $V(r)$, and it computes and displays the other two functions. Finally, when the *two*-dimensional plots are being displayed the program allows the user to enter a "comparison function," which creates a sequence of x's. If you have entered the correct analytic functional form for $\rho(r)$, $V(r)$, or $E(r)$, the x's should lie on one of those curves. Be sure it lies on the right one! Note that the comparison function option is not available when 3-D plots are being displayed.

3.3 Running POISSON

3.3.1 Introduction

The POISSON program is a much more sophisticated program than the GAUSS program. POISSON can deal with any assemblage of 2-D objects inside a grounded box, assuming the objects can be made up from points, lines, rectangles, circles and ellipses. The objects can either be conductors, point charges, or objects whose exterior surface has a potential that varies with position according to some function that you specify. The two-dimensionality of all objects means that they are assumed to have infinite extent in the third dimension. Thus, what appears on the screen as "point" charges are actually infinitely long line charges, etc. In order to remind the reader of this dichotomy, we shall henceforth refer to such objects as "point (line) charges."

The program requires you to use a mouse to draw the system you wish to study. It then proceeds to calculate the potential on a grid of points inside the screen area, using the discretized form of Poisson's equation (Eq. 3.9). When you click on the **Run** hot key, the calculation is performed for the system you have created.

The calculation of the potential at all grid points is done in an iterative manner, i.e., the program begins by assuming starting (zero) values of the potential, except at points having some specified value, and then sweeps through the grid many times, applying Eq. 3.9 at each grid point to get continually updated values. Each sweep through the grid the program checks the convergence of the calculation by comparing the root mean square change from one iteration to the next with some tolerance parameter. Once a solution for the potential is found, the program also calculates the gradient and divergence of the potential. (The charge density at each point in space is the negative of the divergence.) Be aware that the charge density, obtained from numerical second derivatives of the calculated discretized potential, shows large unphysical fluctuations, but its display still contains some useful information when interpreted with this limitation in mind.

A number of displays may be selected: contour maps and 3-D plots for the potential, electric field vectors, a 3-D plot of the charge density, or all four plots at once. In addition, the program allows you to draw electric field lines through any given point selected by mouse, and it also display plots of the potential, electric field, and divergence on different cross sections that you can draw by mouse.

3.3.2 Program Parameters That Can Be Modified

You can easily change the program parameters if their default settings do not suit your task. All the options listed are under the **Control** menu:

- **Exact Solution Option** under **Algorithm and Boundary Conditions**. If you have created a system consisting only of point (line) charges, then it can be solved either exactly, by Coulomb's law, or approximately, by the relaxation method applied to Poisson's equation. One difference between the two methods lies in the choice of boundary conditions at the edges of the screen: grounded box for Poisson's equation, and no boundaries for the exact solution. You can choose which method you want by selecting the **Exact Solution** or **Relaxation**

Algorithm option. **Exact Solution,** means the program uses Coulomb's law[1] to find the electrical potential for point (line) charges. If the **Exact Solution,** has not been selected then the program uses the relaxation algorithm for solving Poisson's equation. You can switch between the two solution methods by clicking one of two buttons on the input screen. Remember that if any conductors are present in the system, the program cannot use the exact solution.

- **Radius of "Point" (Line) Charges** under **Algorithm and Boundary Conditions.** If you are using the exact solution method (Coulomb's law), to find the potential and field of a collection of line charges, the program is actually working with narrow cylinders of charge, so as to avoid singularities. You can specify the radius of these cylinders in terms of the grid spacing. Thus, a row of cylinders of radius $\frac{1}{2}$ lying on adjacent grid points would just be in contact. The cylinders have a uniform charge density of radius R, so that, apart from an overall multiplicative constant, the electrical potential outside them can be expressed as $-\frac{1}{2}\rho \ln r/R$, and inside them, the potential can be expressed as $\frac{1}{4}\rho(1 - (r/R)^2)$. Note that this program feature is only of importance for systems consisting exclusively of point (line) charges, which can either be solved exactly by Coulomb's law or iteratively using Poisson's equation—and the R parameter is not used in the latter case.

- **Relaxation Parameters.** A number of parameters are used to control the relaxation method for solving Poisson's Equation. These may be varied by selecting **Relaxation Parameters** under **Control.** The parameters include—

  - **Relaxation Parameter/Omega.** Values greater than one correspond to "over-relaxation."

  - **Tolerance.** This number defines how closely the solution must satisfy Poisson's equation. Specifically, it is the rms percentage difference between the quantities on both sides of Poisson's equation averaged over the grid. If the tolerance is set to zero, the algorithm would continue iterating until Poisson's equation was exactly satisfied, or until the number of iterations is bigger than **Maximum Number of Iterations.**

  - **Number of Iterations.** You can independently set the maximum and minimum number of iterations to be used. Clearly, if you want to have the program use some specific number of iterations, these two values should be equal.

- **Graphics Parameters**

  - **Number of Iterations Between Displayed Iterations.** Set this to some small number only if you want to see how the relaxation method works, but otherwise keep it high, so as not to slow down the program.

  - **Number of Contour Lines.** The fewer contour lines you have, the faster they will be drawn, but the plot will be less readable.

---

[1]Although we refer to the exact solution method as Coulomb's Law, we actually use the equation for an infinite *line* of charge, not a point.

- **Color Key.** This setting controls whether a color key is displayed that shows the numerical values for the contour lines on the potential plot.

- **Grid Matrix Size.** A larger matrix (finer grid) will give you more accuracy, but it will slow down the calculations. You only need select the number of columns, not the number of rows in the grid. The number of rows is selected automatically, so as to keep the vertical and horizontal grid spacings the same. For many exercises it is important to use the finest grid available (largest matrix), in order to get enough accuracy. Be aware, however, that finer grids may require more iterations to converge properly.

- **"Show Grid" Option.** This setting is used to turn the display of grid points on and off. It is convenient to leave the grid points on when you want to draw objects having a specific size and placement.

- **"Draw With Fixed Center."** This option is used to change the way figures are drawn by the mouse. If it is on, all figures will be drawn from the center outwards (useful if you want to control the placement of the center position of a particular shape). If it is off, figures are drawn from one corner to the opposite corner (useful if you want to control the placement of the starting and ending points of a figure). Squares and circles are much easier to draw with a fixed center, whereas line segments may be easier to draw if the **Fixed Center** option is turned off.

- **Potential/Charge Function.** This option is used when you want to define non-conducting shapes that have a potential or charge distribution that varies with position according to some function that you specify. If the **Potential/Charge Function** option is checked, each time you draw a new shape, you will be asked to enter a distribution function, expressed in terms of the variable t. The convention for lines is that $t = 0$ at the start of the line, and $t = 1$ at the end. For circles, ellipses, and squares, $t = 0$ at zero degrees, and $t = 1$ at 360 degrees. For example, if you wanted to specify that the potential alternated around a circle between $+100$ V and -100 V, with the first quadrant positive, the next negative, etc., you might enter the function: $200 * h(sin(4 * pi * t)) - 100$. (Note that $h(x)$ is the unit step function, which is 1.0, only when its argument is positive, and zero, otherwise. If you just want to specify constant potentials on each object (assuming they are conductors), you should turn off the **Potential/Charge Function** option. Alternatively, you could just enter a constant, when the function is requested. The same option is used to specify the functional dependence with position of the charges for a row of point (line) charges. For example, if you wanted to specify that a row of 21 point (line) charges have charges alternating between $+1$ and -1, you could specify the function $cos(20*pi*t)$. One additional feature of the **Potential/Charge Function** option allows you to treat situations involving gaps in a conductor. For example, suppose you wanted to create a charged conductor in the shape of a semicircle (actually, a long half-cylinder), at a potential of 100 V. The way to do this would be to enter the function $100 * h(sin(2 * pi * t)) + o$. Note that the appearance of the letter o tells the program to "erase" the conductor (here assumed to be a circle), where the entered function has the numerical value of the constant o, which it will for t between $\frac{1}{2}$ and 1, i.e., the bottom half of the conductor.

3.3.3   How To Create a New System

● Choose the type of the first object in the system from the group of icons on the left side of the screen, and set its potential or charge on the slider below. The possible choices of objects include plate, two plates, square, circle (ellipse), point (line) charge, or row of point (line) charges. If you select two plates, they are assumed to have equal and opposite potentials—i.e., they are a capacitor.

● Set the slider bar on the left of the screen to the potential you wish to set the figure you are about to draw. (The slider value is neglected if you have activated the **Potential/Charge Function** option—see last section.)

● Click and drag the mouse to draw the figure you have selected. Release the mouse button to finish the drawing. If you release the button at the point you started at, or somewhere outside the drawing area, the figure will be canceled. The figure will also be cancelled if any part of it is outside the drawing area.

● Repeat the last three operations until you have created as many objects as you wish to use.

● If you want to delete the system you have created and start over, just click on the delete key at the bottom.

● If you want to modify some portion of the system, you can use the **Modify** option. Specifically, this option allows you to change or read the potential of any objects, or redraw, move, or erase them. Simply click on the **Modify** icon. This will allow you the following options:

   – **Select a Figure.** Click on any edge using the left mouse button.

   – **Select a Group of Figures.** Select them one by one using right mouse button, or enclose them in the "rubber rectangle."

   – **Change the Potential or Charge of a Figure.** Select it and then change its value on the slider.

   – **Reshape Figure.** Select the figure and then click and drag the highlighted points.

   – **Move Figure.** Select the figure, and click and drag any edge.

   – **Remove Figure.** Select the figure, and press the **Remove** hot key (F7).

   – **Define or Change the Charge or Potential Function.** Select the figure, and press the **Function** hot key (F8).

   – **To leave modify mode, click the modify icon.**

3.3.4   Calculation and Display

Press the **Run** hot key to have the program calculate the potential on an x-y grid. Note that because of discretization some smooth shapes are approximated by an irregular boundary. For some shapes (plates and rows of charges), discretization effects can be minimized by suitable choice of orientation. Once the potential is calculated, you can investigate the properties of the solution by—

- **Drawing the potential, electric field, or the charge density at all points in the plane.** The potential can be displayed either as a contour map (equipotentials), or as a 3-D surface. The electric field can be displayed using little colored arrows, or as a 3-D plot of its magnitude. Finally, the charge density (computed from $-\nabla^2 V$), can be displayed as a 3-D surface. Note that the charge density plot will show up both charges that you created, as well as induced charges on conductors. All these options are selected under the **Plot How** menu. Since the charge density depends on second derivatives of V, it will be a much less accurate plot than for V itself. *Consequently, in applications where you are interested in observing the charge density, you might want to use the finest grid available.*

- **Displaying cross sections of the potential, electric field, and charge density along any line you draw.** Simply choose the **Cross Section of E and V** option in the **Extras** menu. The program then switches screens, and it allows you to measure the potential, charge, electric field, and charge density at any point on the graph. It will also let you plot graphs of potential and electric field on lines of cross section you choose by drawing with the mouse. (You will need to draw a line longer than about four grid points.) *Note that the plots are color-coded with the numerical display.*

- **Drawing an electric field line through a point.** Simply choose the **Field Line Through a Point** option in **Extras** menu to go into the part of the program that will draw electric field lines through any point on which you click the mouse. By clicking at many points, you can build up a field pattern. However, the density of lines at any point in space is not a reliable indicator of the field strength, in this case. (Note that the program beeps at you if you try to start a field line at places where it is not meaningful, for example, if the potential is constant at all nearby grid points.)

- **Modifying the system.** This will allow you to change any parameters, add new figures, modify old ones, or press the **Erase** hot key to start a new system. At any point you can save your current plate configuration, or read some configuration from a file, by using options from the **File** menu. This option is particularly useful if you have created a system that has taken some time to put together, and you wish to display it later.

- **Comparing the solution to known analytic functions.** The solution found from iterating Poisson's equation can be visually compared to any analytical expressions, or to expressions that can be written in terms of two variables as an infinite series with known coefficients. This option, **Input potential,** appears under the **Extras** menu item. Specifically, the two choices are to input a potential function in rectangular or cylindrical coordinates. The program then creates a contour map of the function you supply that may visually be compared with the solution from Poisson's equation.

3.4 Exercises

The first five exercises apply to the GAUSS program; the remaining ones apply to the POISSON program.

3.1 Uniform Cylinder of Charge

a. Using Gauss' law, prove that the electric field as a function of distance, r, from the axis of an infinite cylinder of radius R filled with a uniform charge density ρ can be expressed as

$$E = \rho r / 2\epsilon_0 \text{ for } r < R, \text{ and } E = \rho R^2 / 2\epsilon_0 r, \text{ for } r \geq R.$$

b. Using the previous expressions for the electric field, prove that the electric potential of a uniform infinite cylinder of charge can be expressed as

$$V = -\frac{\rho R^2}{2\epsilon_0} \ln(r/R), \quad \text{for } r \geq R,$$

and

$$V = \frac{\rho}{4\epsilon_0}(R^2 - r^2), \quad \text{for } r < R.$$

c. If we assume that the charge density ρ is such that $\frac{\rho R}{2\epsilon_0} = 1$, show that the electric field and the potential can be expressed in terms of the variable $x = r/R$, and the unit step function $h(x)$ as

$$E = xh(1 - x) + h(x - 1)/x,$$

and

$$V = \frac{R}{2}(1 - x^2)h(1 - x) - R \ln x(h(x - 1)).$$

d. Use the GAUSS program to find the electric field and the potential for a uniform cylinder of charge of radius $R = 5$. Hint: The appropriate charge density function to enter is simply $h(5 - r)$. See if you can match the results of the program using the appropriate comparison function, and the formulas given in part c.

3.2 Non-Uniform Charge Distributions for Sphere, Cylinder, and Slab

a. Try several functional forms for $\rho(r)$, and calculate by hand what the electric field and potential would be both inside and outside a sphere, cylinder, or slab with that $\rho(r)$. For example, you might try $\rho(r) = r$, and $\rho(r) = R - r$ for $r < R$, and $\rho(r) = 0$ for $r \geq R$. Verify that your solutions for $E(r)$ and $V(r)$ match at $r = R$.

b. For the sphere, cylinder, and slab, enter your $\rho(r)$ functions in the GAUSS program, and see if the solutions you found by hand for $E(r)$ and $V(r)$ match those found by the program. In order to test this, you first need to convert your hand solutions into a single function that holds for all r, using the appropriate combination of $h(5 - r)$ and $h(r - 5)$. Then, you need to enter this function using the comparison function feature of the program. (R must be given a specific numerial value when using this feature.)

3.3 Spheres, Cylinders, and Slabs Having Zero Net Charge

a. Consider functional forms for $\rho(r)$ that change sign. Two of particular interest are $\rho_1(r) = sin(2\pi kr)/r^n$, and $\rho_2(r) = sgn(sin(2\pi kr))/r^n$, where k and n are integers, and $sgn(x)$ is either $+1$ or -1, depending on the sign of x. Try to choose values of k and n for each $\rho(r)$ that will yield zero net charge inside a sphere of radius $R = 5$.

b. Use the GAUSS program to create a sphere of charge that has the charge density function $\rho_1(r)$ inside, and no charge outside. Do this by entering the appropriate combination of step functions: $h(5 - r)$ and $h(r - 5)$ for $\rho(r)$. If you have chosen k and n properly in part a, the net electric charge inside the sphere will be zero. What condition must be satisfied for $E(r)$ or $V(r)$ outside the sphere if the net charge inside the sphere is zero? Does the result for the charge distribution you tried satisfy this condition? (If not, try other values of k and n.)

c. Try to come up with other values of k and n that give zero net charge inside the sphere using the form for $\rho_1(r)$, and see if they do in fact produce that result when GAUSS is run. Also, try using the functional form $\rho_2(r)$ with those same values of k and n.

d. Switch the program to planar symmetry (a thick slab), and again try to come up with values of k and n for which the functional forms for $\rho_1(r)$ and $\rho_2(r)$ have zero net charge inside some region bounded by $(0, +5)$. Express the charge density in terms of the appropriate combination of step functions $h(5 - r)$ and $h(r - 5)$, so that it holds for all r. Enter the charge density function in GAUSS, and see if the results are consistent with zero net charge inside the slab.

3.4 Concentric Thick Shells of Charge

a. Consider a sphere of charge that satisfies $\rho(r) = a$ for $r < R/2$, and $\rho(r) = b$ for $R/2 \leq r \leq R$. Find the electric field $E(r)$ for all r, using Gauss' law by hand.

b. Show that the electric field will be zero outside the sphere of charge if $a = -7b$.

c. Use the GAUSS program with the $\rho(r)$ defined in part a, and see if the electric field is zero outside the sphere when b $= 1$, and $a = -7$. If we assume that the radius of the sphere is 5 units, you could express $\rho(r)$ in terms of the appropriate combination of the three step functions: $h(r - 2.5)$, $h(r - 5)$, and $h(5 - r)$.

3.5   Screened Coulomb Potential

The screened Coulomb potential of the form $V(r) = \frac{q}{4\pi\epsilon_0 r} e^{r/R}$, commonly occurs in a conducting medium.

    a. Derive expressions for the electric field $E(r)$ and the charge density $\rho(r)$, given the above $V(r)$.

    b. Using the GAUSS program set for spherical symmetry, enter the functional form of $V(r)$, and see what $\rho(r)$ and $E(r)$ it produces. For convenience, let the constant $\frac{1}{4\pi\epsilon_0} = 1$, and try various values of R. See if you can match the program's results for $\rho(r)$ and $E(r)$ using the comparison function.

The following exercises apply to the POISSON Program:

3.6   Comparison of Exact and Approximate Solutions

The only case for which the POISSON program can compute "exact" solutions is for point (line) charges in the absence of conductors. In this case, the program assumes that no conditions are imposed on the potential at the edges of the containing box. In contrast, for the approximate solution, based on iterating Poisson's equation, it is assumed the potential is zero at the edges of the containing box. In this exercise, we want to examine the differences between the "exact" and approximate solutions for three charge distributions: a point (line) charge (monopole), dipole, and quadrupole.

    a. Using POISSON, put a point (line) charge at the center of the screen, and examine the exact and approximate solutions. Both the contour (equipotential) and 3-D plot of the potential should be examined in each case. Near the middle of the box the two solutions should look quite similar, but you should observe differences near the boundary— explain these. (The 3-D plot may be particularly helpful.)

    b. You may also observe differences in how steep the central spike at the charge is for the exact and approximate solutions. These differences arise because for the exact solution, the program actually works with a finite cylinder of charge, rather than a point (line), so as to avoid infinities. Try varying the size of the radius of "point" (line) charges (under the **control** menu), between its allowed limits, and see what effect it has on the 3-D plot of the potential for the exact solution. Also, observe cross sections of the potential (under **extras** menu), and see how the potential varies along a line drawn through the "point" charge. What functional form would you expect the potential to have close the "point" charge when its radius is increased to the maximum value allowed? Does that seem to be the case, based on the appearance of the graph?

    c. While in the cross section mode, observe and record the greatest potential found at any point on the boundary of the box for the exact solution. (For the approximate solution, using Poisson's equation, the boundary should have a constant potential of zero.)

d. Now create a dipole—a pair of equal and opposite charges located symmetrically near the center of the box at $x = -a$ and $x = +a$. Record the value you use for a. Observe both the exact and approximate solutions for the dipole using the contour and 3-D plots of the potential and particularly note differences near the boundary of the box. Repeat part c for the dipole. You should find that for the exact solution, the greatest potential at any point on the boundary is much closer to zero (exact and approximate solutions agree better), than for the monopole case. Why?

e. Now create a quadrupole by placing positive charges at (a, a) and $(-a, -a)$ and equal magnitude negative charges at $(a, -a)$ and $(-a, a)$. Make the same observations you did in part d. Explain why the maximum difference between the approximate and exact solutions you find at the boundary of the box is greatest for the monopole, and least for the quadrupole.

f. Repeat parts c–e using twice the value of the constant a, and explain why the differences between the exact and approximate solutions are greater than before.

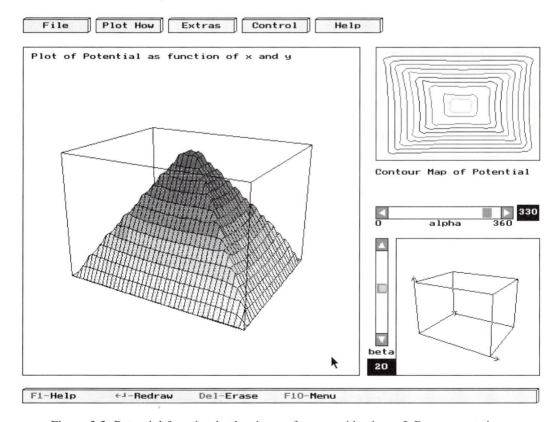

Figure 3.2: Potential function in the shape of a pyramid using a 3-D representation.

3.7   The "Pyramid" Potential

a. Try to figure out what arrangement of point (line) charges or conducting plates produced the pyramid-shaped potential of Figure 3.2. Run the program using this configuration, and see if you get a pyramid when using the 3-D representation.

b. Give an argument why the potential should have a pyramid shape for this particular situation. Hint: Consider the potential variation along the edge of the pyramid, and show that it would be expected to satisfy Poisson's equation.

3.8   Local Maxima or Minima in the Potential

a. A function of two variables, such as $V(x,y)$, has a local maximum or minimum only where its value is greater (or less) than the average of its value at all nearby points. Use this fact, and the discretized form of Poisson's equation, to show that the potential can only have a maximum or minimum at places where charge is located.

b. Use the POISSON program to create a rectangular conductor at a potential of 100 volts at the center of the display area, and show the equipotential contours and 3-D plot of the potential. Explain why you would expect to find the potential inside the rectangle to be constant. Is it?

c. If not enough iterations are made, you may find that the potential inside the rectangle is not constant—as evidenced by the existence of equipotential contours inside it. (Such contours imply the existence of a maximum or minimum in the potential inside the box.) Why does the presence of an interior equipotential contour imply a maximum or minimum? Try varying the number of iterations, and see the effect this has on such non-physical contours. You can easily force the program to use some specific number of iterations, by setting the maximum and minimum number of iterations equal to the same value.

d. For a better probe of whether the potential has a maximum or minimum inside a rectangular conductor look at cross sections of the potential under the **Extras** menu, and find the potential numerically at various points inside the box by positioning the mouse. Record the potential, $V(0, 0)$, you observe at the center of the rectangle when the program is run with varying number of iterations. Make a plot of $(V(0, 0) - 100)$ versus the number of iterations, and comment on the plot.

e. Again record $V(0, 0)$, but this time restore the default values of the minimum and maximum number of iterations, and instead see how $V(0, 0)$ varies as you change the "tolerance" parameter. Make a plot of $(V(0, 0) - 100)$ versus the tolerance parameter.

3.9   Saddle Points

The potential in a charge-free region of space cannot have a local maximum or minimum, as noted in the previous exercise. It can, however, have saddle points, which show a maximum in one direction, and a minimum along a perpendicular direction. One example of a potential with a saddle point is the potential of a quadrupole.

  a. Create a quadrupole with the POISSON program by placing positive charges at (a, a) and ($-a$, $-a$), and equal magnitude negative charges at (a, $-a$) and ($-a$, a), and display the equipotential contours and the 3-D plot of the potential. Where is the saddle point located?

  b. Go to the cross section mode, and display the potential along a diagonal line connecting the two positive charges. Do the same along a line connecting the two negative charges. Along which diagonal is there a minimum, and along which is there a minimum at the saddle point? What is the value observed for the electric field at the saddle point? Why must it have that value? What does the existence of a saddle point imply about the stability of a free electron placed at that point?

3.10 Over–Relaxation Parameter

The value of the over-relaxation parameter, ω, appearing in the iterative solution to a discretized form of Poisson's equation determines the stability and the rate of convergence of the solution. Specifically, the larger we make ω, the faster the convergence, but the more unstable the solution—meaning that it begins to diverge as the number of iterations increases. In this exercise, we investigate the trade-off between stability and rate of convergence.

  a. Create a conducting cylinder (circle) at a potential of 100 volts placed at the center of the screen. Set the value of ω to 1 (no over-relaxation), and the number of iterations (both maximum and minimum) to 10. Display the equipotential plot. Are there some contours inside the cylinder, indicating that the solution has not yet converged? Switch to cross section mode, and measure and record $V(0, 0)$, the potential at the center.

  b. Repeat step a when the number of iterations, N, is set to 20, 30, 40, 50, and 60. Make a graph (by hand) of $V(0, 0)$ versus N, and label the curve $\omega = 1$.

  c. Change ω to 1.25, and generate another set of data for varying number of iterations. Plot the data on the same graph, labeling the graph $\omega = 1.25$. Do the same for values of $\omega = 1.5$, 1.75, and 2.00.

  d. Based on the shapes of the curves you plotted, what do you conclude about the way the rate of convergence and the onset of instability varies with the parameter ω? What value of the parameter appears to be the best choice? (Some sources suggest the value $\omega = 1.5$ represents a good trade-off between stability and rate of convergence.)

3.11 Electrostatic Shielding and Conductors

  a. Place a point (line) charge inside a cylindrical conductor, and observe the equipotential pattern for several locations of the point (line) charge.

(The 3-D plot of the potential is also highly instructive.) Does the equipotential pattern and electric field outside the conductor appear to depend on where the interior point (line) charge is located? (Note that because the conductor is created to have a fixed potential, the equipotential pattern outside it is actually the same with or without the charge placed inside. This is a non-trivial result, because the charge on the face of the conductor needs to redistribute itself depending on the placement of the interior point [line] charge.)

b. Place a second point (line) charge outside the conductor. Keep the inside charge location fixed, while you try several locations for the outside charge. Does the interial pattern of equipotentials (or 3-D surface) change in any way as the outside charge is moved from place to place? (You may wish to repeat these observations using other conducting shapes, such as a rectangle.)

3.12   Creating a "Channel"-Shaped Conductor

Consider a charged hollow conductor that has an opening, so that it does not completely shield the interior from any outside fields. We wish to investigate, in the next exercise, how the extent of shielding depends on the size of the opening, relative to the size of the conductor, but first, we shall see here how POISSON can create a hollow conductor using three different methods. Consider a rectangular enclosure with its right side missing, which we shall describe as a "channel." Here are two ways to create a channel charged to 100 volts. Try them both, and see if they produce the same equipotential and 3-D plots.

a. Draw three connected line segments (plates) set to 100 volts to form a three-sided channel.

b. Activate the **Charge/Potential Function** option under **Control**, and draw a rectangle. Enter the function: $100 * h(abs(t - 0.5) - R/2)) + o$, where R is the exact ratio of the height of the rectangle to its width, and o is the letter o, not the number zero. This function has the property that its value is 100 for all values of $|t| > R/2$, i.e., on the three sides of the rectangle other than the right-most one. It also has the value o on the right-most side—which is a signal to the program to remove that side. Note, that after "removing" one side of the rectangle the drawing is unfortunately not changed, but you can tell it has been removed from the 3-D plot of the potential. If you use a value for R in the preceding equation, which is half the rectangle's height-to-width ratio, notice its effect on the 3-D plot of the potential. What conducting shape have you created with this value of R?

3.13   Shielding in a Partially Open Enclosure

Consider a "channel"-shaped hollow conductor—a rectangular box with one side missing (see previous exercise). We wish to investigate how the extent of shielding inside the channel depends on the ratio of its height to its length, i.e., its "aspect ratio," R.

a. Create a channel-shaped conductor with $R = 1$ at a potential of 100 volts. Observe the equipotential contours and the 3-D plot. Switch to the cross section mode (under **Extras**), and measure the electric field at the midpoint of the open side of the conductor, E_0. Then, observe how far in from the opening you need to move for the strength of the electric field to fall to $0.1\ E_0$. Divide this distance by the channel width along the x-axis, and call the result x.

b. Repeat step a for channel-shaped conductors having aspect ratios of $R = 0.2, 0.4, 0.6$, and 0.8, and record x in each case. Make a plot (by hand) of x versus R. What do you conclude from the plot about how the extent of shielding inside a partially open conductor depends on the conductor's aspect ratio? Explain why your finding is reasonable.

3.14   Image Charges

Consider the problem of a point (line) charge, q, located a distance r from the center of a grounded $(V = 0)$ cylindrical conductor of radius R. According to the method of image charges, the electric field everywhere outside the cylinder is the same if the conducting cylinder is replaced by an appropriate "image charge." (For more details on the method of image charges, see chapter 4.)

a. Use the method of image charges to calculate (by hand), the appropriate value and position of the image charge for arbitrary q, r, and R. Note that we are dealing with conducting cylinders here, not spheres.

b. Use the POISSON program to create a grounded cylinder $(V = 0)$, and a point (line) charge some distance away. Record the values you use for q, r, and R, and calculate the appropriate image charge and location from the expression in part a. Observe the equipotential contours and 3-D plot for the cylinder plus point (line) charge.

c. Remove the cylinder by going into "modify mode," clicking the mouse on the cylinder, selecting the **Charge/Potential Function**, and entering the letter o. (The reason for removing the cylinder by this relatively complicated method is that the cylinder has been removed from the system [as you can verify from the equipotential plot], but its drawing will remain on the screen, which is an advantage in being able to explore the equipotential plot outside the region where the cylinder used to be.) Having removed the cylinder, now replace it by an image charge of the predicted value located at the predicted location. See if the equipotential plot outside the former cylinder looks just like the original plot of cylinder plus point (line) charge—as it should if your image charge has the correct value and location.

d. Explain why the method of image charges works. Hint: Make various equipotential plots for various combinations of two unequal point (line) charges having opposite signs. Observe in each case the shape of the equipotential contour for $V = 0$.

3.15   Approximating a Finite Sheet of Charge

Consider a uniform sheet of charge of width $2a$ which lies between $x = -a$, and $x = +a$, and is infinite in length along the z-axis—perpendicular to the screen.

a. Derive exact formulas for the electric field and potential at any point on the y-axis. Make a plot of $V(y)$ versus y by explicitly finding the values for $y/a = 0, 0.2, 0.4, \ldots, 1.0$, and connecting the points with a curve.

b. Approximate the sheet of charge using a row of equally spaced point (line) charges placed along the x-axis centered on the origin—use the icon for a row of charges. Fill up the middle third of the screen with the row of charges, and use the maximum number of charges allowed by the program.

c. Find the equipotential plot, using the "exact solution" method (under **Control—Algorithms and Boundary Conditions**), and switch to the cross section mode. Measure the field and the potential at points along the y-axis, specifically: $y/a = 0, 0.2, 0.4, \ldots, 1.0$. Compare the values you find with the values you compute using the formula you derived in step a, by plotting the values on the same graph you created in part a. Remember that to get agreement, you need to use a common charge in both cases. Which y-values, small or large, show the poorest agreement between the two curves? Why is that?

d. Repeat steps c and d using a row of charges containing fewer charges than before, but having the same length, and comment on the results.

3.16   Charge Distribution Along a Straight Conductor

a. Using the finest grid size, create a straight conductor (plate) at the center of the screen at 100 volts. Display the equipotential plot and the 3-D plot of the potential—comment on its shape. Also, display the 3-D plot of the charge distribution—comment on its shape. Switch to cross section mode, and estimate roughly how far you need to be from the ends of the conductor for the charge density (the divergence of V), to fall to $1/e = 68\%$ of its value at the ends. Express that distance as a fraction of the length of the line.

b. Create a row of electric charges the same length as the conductor you created in part a, and display the equipotential plot and the 3-D plot of the potential. In what specific respect do the plots differ from what was found for a straight conductor? Try various numbers of charges.

c. Using the largest number of charges available, select the **Charge/ Potential Function**, and see if you can come up with a function that results in equipotential curves that closely resemble those found for the straight conductor. One promising possibility might be $exp(-a * (abs(t - 0.5) - 0.5))$, for various values of a. Note that this function puts the maximum charge density at the conductor's ends ($t = 0$ and

1), and the charge density drops to $1/e$ of its value a distance a from each end. (See part a for a promising value of the constant a.)

3.17   Fringing Field of a Capacitor

a. Create a capacitor at the center of the screen. In order to get an accurate display of the equipotentials, you may want to use the finest grid available—set grid size under **Control** menu item. After finding the equipotential contours, switch to the **Extras** and draw field lines through a number of points both inside and outside the capacitor. What happens to the field as you approach the edges of the capacitor from the inside?

b. Create a series of capacitors whose plate separation is kept fixed at, say, a quarter the screen height, and whose plate length covers a range of values. Let the ratio, r, of plate length to plate separation take on values $r = 1, 2, 3, 4,$ and 5. For each of the five capacitors, after you display the equipotentials, switch to cross section mode, and see how the field varies along a horizontal line midway between the plates. On the line measure the field at the very center of the capacitor, and also the field at its edge. Find and record the ratio, R, of these two fields for each capacitor.

c. Make a plot (by hand) of R versus r. What do you conclude from this plot about how the fringing field of a capacitor depends on its shape?

3.18   Creating an Enclosure With a Uniform Electric Field

Using the finest grid size, create a capacitor with large plate separation in the center of the screen charged to 100 volts. As you noticed in the previous exercise, the electric field strength decreases as you move towards the edges of the plates from the inside. It is possible, however, to make the electric field inside the capacitor entirely uniform by adding insulating plates on its two sides. The potential must vary as a function of position along these plates so that (a) it is a linear variation, and (b) it matches the potential at the upper and lower plates. Try to come up with the appropriate function. Set the **Charge/Potential Function** option under **Control**, and add the two side plates having the desired potential function. Observe the equipotential plot, and the 3-D plot of the potential. Also, draw field lines through many points to show the field pattern inside and outside the capacitor. It should be obvious from the result whether you chose the correct potential function for the side plates. Can you explain why the particular function you chose forces the electric field to be exactly uniform inside the capacitor?

3.19   Conducting Cylinder in a Uniform Electric Field

a. Using the finest grid, select the boundary condition to be a parallel plate capacitor. Create a cylinder (circle) whose potential is zero at the center of the box. Observe the equipotentials and 3-D plot. How uniform does the field inside appear to be?

b. What is the predicted charge density as a function of angle for a conducting grounded cylinder placed midway between plates charged to +100 volts and −100 volts? What is predicted for the net charge on the cylinder?

c. Observe the 3-D plot of the charge density. What would you expect this plot to look like, given the functional dependence of charge density on angle you suggested in part b? Discuss the reasons that the plot does not look quite like what you might predict. Does the plot appear to be consistent with there being zero net charge? In what respect?

d. Now remove the cylinder from the plot. Predict and record the potential difference between adjacent horizontal equipotential lines, given that the top and bottom of the box are at +100 and −100 volts. Switch to cross section mode and verify your prediction by moving the mouse.

e. Create a conducting cylinder which is located above the center of the box by an amount equal to the spacing between adjacent equipotential lines. What do you have to set the potential of this cylinder at for it to have zero net charge? Observe whether or not the cylinder does appear to be consistent with having zero net charge, by looking at the 3-D plot of the charge density. Try changing the potential of the cylinder, and report what effect that change has on the 3-D plot of the charge density.

f. The boundary conditions used in creating the box ensure that near its edges the electric field is uniform with or without other objects present. Explain why placing a conducting cylinder in such a box is not quite the same as placing it in a uniform external electric field. Why is the approximation to the uniform external field best when the cylindrical conductor is (a) placed in the center of the box, and (b) grounded ($V = 0$).

3.20 Uniform Field Inside a Cylindrical Shell

a. Consider a mathematical cylindrical surface whose axis is at right angles to a uniform electric field along the x-axis. Let the potential be zero at the axis of the cylinder. Prove that the potential at points on the cylinder surface can be expressed as a function of angle as $V = -V_0 \cos \theta$.

b. Now consider a *real* hollow cylindrical shell made of dielectric material. Suppose we placed charges on its surface in such a way that the potential around the cylinder varied according to $V = -V_0 \cos \theta$. Explain why, based on the uniqueness theorem, the field everywhere inside the cylindrical shell must be uniform.

c. Using the finest grid size, create a cylinder at the center of the screen which occupies the middle third of the screen. Specify that the potential, as a function of angle satisfies $V = -V_0 \cos \theta$. To do this, you need to enter the potential function: $-100 * cos(2 * pi * t)$, assuming a $V_0 = 100$ volts.

d. Display the equipotentials, and create field lines through many points inside the cylinder. To what extent does the field appear to be uniform? To what do you attribute the fact it may not be exactly uniform?

e. Display the 3-D charge density plot, and observe where the charge tends to be concentrated. Can you explain why the charge is concentrated at those four points? (Hint: Use the program to create a quadrupole—two positive charges at $(-a, a)$ and (a, a), and two negatives at $(-a, -a)$ and $(a, -a)$—and observe the electric field near the center of the quadrupole.)

3.21 Pure Multipole Fields

The potential in polar coordinates inside or outside a cylindrical shell can be expressed as a sum of terms of the form $\cos n\theta$ and $\sin n\theta$, specifically

$$V_{ins}(r, \theta) = \frac{1}{2}A_0 + \sum_n r^n(A_n \cos n\theta + B_n \sin n\theta)$$

$$V_{out}(r, \theta) = A_0' \ln r + \sum_n r^{-n}(A_n \cos n\theta + B_n \sin n\theta)$$

Depending on the symmetry of the problem, we may have only the sine or cosine terms present. In this exercise we shall consider dipole ($n = 2$), quadrupole ($n = 3$), and octupole ($n = 4$) terms in this series. Note, that the r and θ functions depend on the same integer n.

a. Create a cylindrical shell at the center of the screen on which the potential has the form of a pure dipole, i.e., $V(\theta) = V_0 \cos 2\theta$. You can do this by entering the potential function, for example, $100 * cos(4 * pi * t)$, when drawing the cylinder. Given this θ dependence, what would you predict for the r-dependence of the potential inside and outside the cylinder?

b. Display the equipotentials, and go to the cross section mode under **Extras** to display several slices through the center of the cylinder. Observe and record the potential at points $r/R = 0, 0.5, 1.0, 1.5, 2.0$. (all at fixed θ), and note whether it has the predicted r-dependence. Make the same observations at another θ.

c. Create a quadrupole cylinder, $V(\theta) = \cos 4\theta$, and repeat parts a and b. Do the same for an octupole cylinder.

d. For which of the following cases—dipole, quadrupole, and octupole—is the agreement between the predicted and observed radial dependence of the potential best? Why do you suppose that is?

3.22 Split Cylinder

Consider a cylinder split in half, so that the upper half ($\theta < \pi$) is at a potential $+V_0$ and the lower half is at a potential $-V_0$. The solution for the potential inside and outside the cylinder can be expressed in the form of an infinite sum of the form given in the previous exercise.

a. Explain why all the $A_n = 0$ for this problem. (Different reasons apply to $n = 0$, and higher coefficients.)

b. For the inside solution, try to prove that the coefficients B_n vanish for even n, and for odd n are given by $B_n = 2V_0/(\pi n R^n)$. To do this you need to match the inside solution on the cylinder to the value V_0 for $\theta < \pi$. The technique is similar to that used to derive coefficients of terms in a Fourier Series, if you are familiar with that.

c. Use the program to create a split cylinder using the potential function $100 * sgn(sin(2 * pi * t))$. Note that because of the *sgn* function, the cylinder will have a potential $+100$ volts on its upper half ($t < \frac{1}{2}$), and -100 volts on its lower half. Can you come up with another function using the step function $h(x)$ that would work just as well? Display the equipotentials and the 3-D plot for the split cylinder. (The 3-D plot can also be used to check that you, in fact, have created a split cylinder.)

d. Use the **Input Potential** feature (under **Extras**) to see if the series solution agrees with that obtained from the POISSON program. For a cylinder of unit radius ($R = 1$), apart from an overall multiplicative constant, the formula given in part b implies $B_n = \frac{1}{n}$, so that

$$V(r, \theta) = \sum_{odd\ n} \frac{r^n}{n} \sin n\theta,$$

where the sum extends over odd n values only. Try this expression in the **Input Potential** function, using different number of terms in the series, and visually compare the resulting display with the solution from Poisson's equation. The region with $r < 1$ is inside the cylinder. *You probably will not get meaningful results unless you suppress the function outside of the unit radius cylinder, because the series terms diverge for large* n. If you want to display the solutions inside the cylinder only, simply multiply each term in the above series by the unit step function, $h(1 - r)$. You could also compare the series solution *outside* the split cylinder ($r > R$). The only difference from the preceding formula is that you need to use negative integral powers of r in the series formula. You should multiply each term by the unit step function, $h(r - 1)$, so as to blank out the solution in the region it does not apply.

3.23　Partially Open Cylindrical Shells

a. Create a half-cylindrical shell at the center of the screen charged to 100 volts. This can be accomplished by entering the potential function: $100*h(0.5 - t) + o$, where it is the letter o that appears in this equation. The *o* tells the program to remove the upper half of the cylinder (even though it remains in the drawing). Without that *o*, we would be specifying the potential to be 0 volts on the upper half and 100 volts on the lower half of the cylinder. Run the program with and without the added *o*, and observe and describe the difference in the equipotential and 3-D plots. (From the 3-D plot, the difference between the two cases should be especially clearcut.)

b. Create cylindrical shells that have less than half missing, using the potential function $100 * h(c - t) + o$, where the constant c takes on values greater than $\frac{1}{2}$. Observe the equipotential and 3-D plots in each case. Also, observe the 3-D plot of the charge density. Explain what you find about how the distribution of charge density changes, as you create cylinders with smaller and smaller missing segments. Also, comment on the way in which the extent of shielding inside the cyinder changes as the size of the missing segment decreases. You may want to switch to cross section mode, and measure and record the electric field at the center of the cylinder, as the size of the missing segment changes. You could then plot the field at the center as a function of the fraction of the cylinder that is missing (the value of $1 - c$).

3.24 Concentric Half-Cylindrical Shells

a. Using the special o variable (see previous exercise), create a conducting half-cylinder centered on the screen at a potential $+100$ volts. Make its diameter about half that of the screen, and make the bottom half missing. Use the 3-D plot of the potential to verify what you created. Create a second conducting half-cylinder concentric with the first, but half the radius of the first, and make its potential -100 volts. (Geometrically, you have two "quonset huts"—one inside the other.)

b. Display the equipotentials and the 3-D plot. Explain why the equipotentials are circular arcs for the middle region between the two half-cylinders, but deviate from circular arcs as you approach the ends of the half-cylinders (on the x-axis).

c. If you were to connect the points on the outer cylinder lying at $\theta = 0$ and $\theta = \pi$ by a line, it is possible to choose a potential variation along that line, such that the equipotentials between the inner and outer upper half-cylinders are exactly circular—a $\ln r$ potential. See if you can figure out what the proper function should be. Try it in the program, and see what equipotential contours it produces.

3.25 Charge Density and Radius of Curvature

a. The charge density on a conductor of convex shape is proportional to the radius of curvature at various points. Give an argument why you would expect this to be true.

b. Using the finest grid, create an ellipsoidal conductor whose major and minor axes are in the ratio 2 to 1, which is charged to 100 volts. Display the equipotentials and the 3-D plot of potential. Also, display the 3-D plot of the charge density. Approximately by what factor is the charge density at the ends of the major axis greater than the ends of the minor axis? Repeat using another ratio, say, 4 to 1, for the major and minor axes.

c. Create ellipsoidal conductors having greater ratios of major to minor axes, and observe the 3-D plot of the charge density. Explain why the distribution of charge density varies in the way you observe. Why does the concentration of charge density at the ends of the major axis stop increasing at a certain point as the ellipses are made more eccentric? Hint: Change the grid size, and see what you get.

3.26   **Sharp Points on Conductors**

The charge density on the surface of a convex charged conductor varies in proportion to the curvature (it is inversely proportional to the radius of curvature), so obviously, charge tends to primarily concentrate on sharp points.

a. Construct a rectangle charged to 100 volts, and display the equipotentials, and the 3-D plot of the potential. Also, display the 3-D plot of the charge density, and observe where the charge tends to be concentrated.

b. Systematically vary the ratio of the length and width of the rectangle, and see what effect this has on the amount of charge at the corners relative to the charge on each side. Also, see what is observed when you change the size of the rectangle, while its ratio of length to width is kept constant. Why do you suppose there appears to be more charge along the rectangle's sides as the rectangles get smaller? (Hint: Try running the program for a rectangle of fixed size using several different grid sizes.)

3.27   **Concave Charged Conductor**

a. Consider a thick cylindrical shell sliced in half along its axis—so that it looks like a "quonset hut" with thick walls. In order to create such a conductor, first make a circle of diameter about half that of the screen by specifying a potential function: $100 * h(0.5 - t) + o$. (Recall, that the o tells the program to remove portion of the conductor, when the potential has this value.) Next, create a concentric circle of half the radius using the same potential function. Finally, draw two line segments (at $V = 100$ volts), joining the inner and outer circles at $\theta = 0$, and $\theta = \pi$. Using the finest grid, display the equipotential plot, and its 3-D plot. Is the result what you expected?

b. Display the 3-D plot of the the charge density. Note where the greatest and least concentrations of charge density appear. Explain why the charge density is much greater on the convex side of the conductor than the concave side, even though the curvature is twice as great in the former case.

3.28   **Three Plates**

a. Place two grounded ($V = 0$) equal-length horizontal conductors above and below the x-axis. Connect the left edges of the plates with a vertical conductor charged to 100 volts. (If this were a real physical system, there would have to be some insulation at the corners separating the plates.) Display the equipotentials and the 3-D plot of the potential. Is it what you expected?

b. It can be shown that inside the region between the horizontal plates, the solution for the potential has the following form:

$$V(x, y) = \frac{4V_0}{\pi} \sum_{odd\ k} \frac{1}{k} e^{-kx} \sin ky .$$

To see how well this series solution agrees with the solution found from the program, enter the preceding series expression in the **Input Potential**. Note that you can ignore the multiplicative constant $4V_0/\pi$, and set the ranges on x and y to be the interval $(0, \pi)$. See how well the result resembles the solution from Poisson's equation as the number of terms in the series is increased.

c. It can be shown that the previous infinite series solution has the following explicit sum:

$$V(x, y) = \frac{2V_0}{\pi} \arctan\left(\frac{\sin y}{\sinh x}\right).$$

See what the display looks like when you enter this function under the **Input Potential** option.

d. Note, that neither the series formula nor its explicit form above can be expected to reproduce the solution *outside* the region between the plates. Verify this statement by displaying the **Input Potential** over the range in x and y in the interval $(-\pi, 2\pi)$. You should find that for regions outside the interval $(0, \pi)$, the "input potential" display bears no resemblance to the solution found in part a.

3.29 **Four Plates**
For a variation on the last exercise, suppose that a second vertical plate (at 100 volts), is added connecting the right edges of the two horizontal grounded plates. Repeat the exercises of the previous problem. It can be shown that the potential inside the region bounded by the four plates can be expressed in terms of the following series solution:

$$V(x, y) = \frac{4V_0}{\pi} \sum_{odd\ k} \frac{\cosh kx \sin ky}{k \cosh k} .$$

3.30 **Charged Square Cylinder**

a. Create a conducting rectangle charged to 100 volts located at the center of the screen. Make its length and width half the screen dimensions in each direction. Display the equipotentials and the 3-D plot.

b. It can be shown, using Fourier series, that a solution to this problem outside the rectangle can be written as the following double sum:

$$V(x, y) = C \sum_{odd\ m} \sum_{odd\ n} \frac{1}{mn} \cos nx \cos my ,$$

where C is a constant that we shall ignore. We can write the nine lowest-order terms as a single sum:

$$V(x, y) = \sum_{odd\ m} \frac{1}{m} \left(\cos x + \frac{1}{3} \cos 3x + \frac{1}{5} \cos 5x \right) \cos my .$$

Enter this function in the **Input Potential**, with limits on x and y both in the range $-\pi$ to π. See how well the results agree with those from solving Poisson's equation. Make a 3-D plot of the input potential. Where is the agreement with the solution from part a poorest?

3.31   Cylindrical Boundary Conditions

Consider the region inside a hollow circular cylinder of unit radius on which the potential has some known functional dependence $V(1, \theta)$. The potential everywhere inside the cylinder can be found in terms of an infinite series that depends on $V(1, \theta)$. Two sample cases would include $V_1(1, \theta) = \theta - \pi$, and $V_2(1, \theta) = \sin^2 \theta$. It can be shown that in the first case, the appropriate series solution is

$$V_1(r, \theta) = 2 \sum_{n=1}^{\infty} \frac{1}{n} (1 - r^n) \sin n\theta,$$

and in the second case we have

$$V_2(r, \theta) = \frac{1}{2} (1 - r^2 \cos 2\theta).$$

a. Create a circular cylinder around which the potential has the functional form specified by $V_1(1, \theta)$. Check that you have created the correct potential function by examining the 3-D plot of the potential, and looking at its value on the cylinder surface.

b. Enter the above form for $V_1(r, \theta)$ using the input potential feature, and see how well the analytic solution reproduces that from Poisson's equation inside the region of the cylinder. (In order to focus your attention on that specific region, you should multiply the above form for $V_1(r, \theta)$ by the step function $h(1 - r)$, so as to make the function zero everywhere outside the unit circle.)

c. Repeat the problem for $V_2(r, \theta)$.

3.5   Structure of the Program

The program consists of three parts: a tools unit, a plates unit, and a main program that calls these other two units, and controls the flow of the program.

The tools unit (POISSONT) consists of a set of physics-related procedures and objects that are used in other parts of the program. All the procedures in this unit can be modified without influencing other parts of the program. They have been written to make them easy to use in other programs. All the code that is worth analyzing or modifing from the physicist's point of view is in the tools unit, unless of course you are interested in the interface or graphics details.

The plates unit (POISSONP) comprises the code responsible for drawing, modifing, and storing objects such as plates and cylinders. To the outside world, it is just a single, independent object that handles all operations in the drawing area and returns initial conditions for iteratively solving Poisson's equation. However, internally it is a link list of objects, each representing one plate or point (line) charge. This structure is managed by another object called TPlates, which is also the only connection to the outside world. *There are no physics-related procedures in this unit.*

Bibliography

1. Gould, H. and Tobochnik, J. *An Introduction to Computer Simulation Methods—Applications to Physical Systems.* Reading, MA: Addison-Wesley Publishing Company, 1988.

2. Lorrain, P. *Electromagnetic Fields and Waves.* 3rd ed. San Francisco: W. H. Freeman & Company, 1988.

3. Reitz, H. R., Milford, F. J., and Christy, R. W. *Foundations of Electromagnetic Theory.* 3rd ed. Reading, MA: Addison-Wesley Publishing Company, 1979.

4. Griffiths, D. *Introduction to Electrodynamics.* 2nd ed. Englewood Cliffs, NJ: Prentice Hall Publishing Company, 1989.

5. Visscher, P. B. *Fields and Electrodynamics.* New York: John Wiley & Sons, 1988.

6. See list of references on the relaxation method in Houtman, H., Jones, F. W., and Kost, C. J., "Laplace and Poisson Equation solution by RELAX3D," Computers in Physics, **8**, 4 (1994) 469–79.

4

Image Charges and the Multipole Expansion

Lyle D. Roelofs
Physics Department, Haverford College

It is the great beauty of our science that advancement in it, whether in a degree great or small, instead of exhausting the subject of research, opens doors to further and more abundant knowledge, overflowing with beauty and utility.

—Michael Faraday

4.1 Introduction

The most fundamental problem in electrostatics is to determine the electric potential $\Phi(\vec{r})$ and field $\vec{E}(\vec{r})$ in all space, given a full specification of all free charges and equipotential surfaces present. Entire fields of mathematics were invented by great mathematicians, Laplace and Gauss among others, to deal with this problem analytically. The Image Charge and Multipole Expansion (IMAG&MUL) simulation illustrates two approaches to this problem that are covered in undergraduate electromagnetism courses. Background material for these simulations is covered here; for further information and detail you can consult any of the standard textbooks in this area, including Griffiths[1]; Reitz, Milford, and Christy[2]; and Corson and Lorain[3], among others.

The electrostatic potential satisfies Laplace's equation,

$$\nabla^2\Phi = 0, \tag{4.1}$$

or, when there is free charge density, $\rho(\vec{r})$, present, Poisson's equation,

$$\nabla^2\Phi = -\frac{\rho(\vec{r})}{\varepsilon_o}, \tag{4.2}$$

where ε_o is the vacuum dielectric constant. If one has solved for $\Phi(\vec{r})$ then $\vec{E}(\vec{r})$ can be readily obtained from

$$\vec{E} = -\vec{\nabla}\Phi. \tag{4.3}$$

It is important to have clearly in mind the conditions that constitute a well-posed problem in electrostatics. One needs to know ρ, of course, if there is any free charge density about, but also since Eqs. 4.1 and 4.2 are second-order, one needs to have information on boundaries fully enclosing the region of interest. Physical insight suggests that the problem is well-posed if Φ, i.e., the voltage, is known on all boundaries, both interior and exterior, or if $\vec{E}$ is known on some enclosing surface. (This hypothesis is proved using Green's theorem in more advanced courses; we will simply assume it.) We will restrict ourselves in the present treatment to the former situation, called *Dirichlet boundary conditions*. Mathematical theorems guarantee the uniqueness of solutions to these problems, so that if a solution is found it is the one and only solution.

4.2 The Method of Images

4.2.1 Introduction

The method of images is a clever and subtle trick that can be used to obtain Φ and $\vec{E}$ in many situations involving a small number of charges and conducting, i.e., equipotential surfaces. The amazing thing about the method of images is that it requires almost no work at all, but is able to address problems that by any other method would be painfully complicated. Some ingenuity, however, is needed.

The method of images is covered in all electricity and magnetism textbooks. Coverage of the technique usually follows closely on the heels of the theorems establishing the uniqueness of solutions of the electrostatic Eqs. 4.1 and 4.2. The method of images exploits the uniqueness idea in a "uniquely" clever fashion most readily seen by way of example.

4.2.2 The Basic Idea and an Example

Figure 4.1 displays a problem in electrostatics. A charge Q is located a distance d from a grounded conducting plane. Φ therefore vanishes everywhere on the conducting surface. The direct approach to this problem is complicated by the fact that the conducting surface responds to the presence of the point charge by developing an induced surface charge density to cancel the field within itself. As this charge density is not known a priori, Eq. 4.2 cannot be applied.

The method of images offers a much easier approach to this problem. Figure 4.2 shows the electric field lines due to point charges $\pm Q$ separated by a distance $2d$. In this case it is no work at all to solve for $\vec{E}$ or Φ. Rather than using Eqs. 4.1 and 4.2 we can write down the solution directly using the known form of the potential of a point charge and the principle of superposition:

$$\Phi(\vec{r}) = \frac{1}{4\pi\varepsilon_o} \sum_i \frac{Q_i}{|\vec{r} - \vec{r}_i|}. \tag{4.4}$$

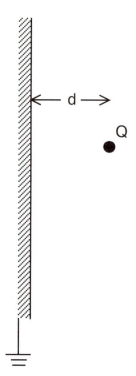

Figure 4.1: Charge Q located near an infinite, grounded conducting plane. Φ can be taken to tend toward 0 far from the charge.

In Eq. 4.4 point charges Q_i are located at positions $\vec{r}_i$. In the present situation Eq. 4.4 gives for the potential

$$\Phi(x, y, z) = \frac{1}{4\pi\varepsilon_o}\left(\frac{Q}{\sqrt{(x-d)^2 + y^2 + z^2}} - \frac{Q}{\sqrt{(x+d)^2 + y^2 + z^2}}\right) \quad (4.5)$$

It is evident from the figure that the plane at $x = 0$, denoted by the dashed line, is an equipotential surface, since by symmetry the electric field is everywhere orthogonal to it. Furthermore, one sees readily from Eq. 4.5 that Φ vanishes at $x = 0$ for all y and z.

We note two interesting facts about this solution for the region $x \geq 0$. First, notice that the free charge distribution is identical to that of Figure 4.1. Secondly note that Φ of Eq. 4.5 satisfies the boundary conditions applicable to the situation depicted there. According then to the uniqueness ideas noted in subsection 4.2.1, the solution written as Eq. 4.5 and depicted in Figure 4.2 is also valid for $x \geq 0$ for the situation depicted in Figure 4.1. It is important to emphasize that the method does not provide the solution in the conducting region, $x < 0$. There the potential is constant (and, for this grounded case, equal to 0) everywhere since $\vec{E}$ vanishes inside a perfect conductor.

The Image Charge mode of the IMAG&MUL simulation is a exploration of this approach. It is suggested that you familiarize yourself with the program and its style of dealing with image charge problems before continuing. Run the program (see section 4.5 for detailed instructions)—it starts by default in the Image Charge mode displaying the point charge and plane situation—examining the

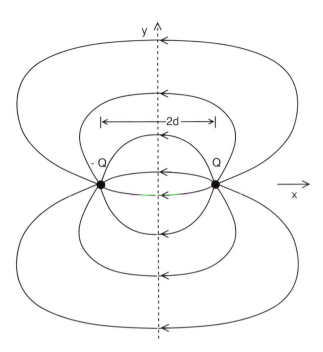

Figure 4.2: Charges $+Q$ and $-Q$ located a distance $2d$ from one another. The lines denote the associated electric field.

display options and adding a negative charge in the correct position to "solve" the problem. Use the *field probe* to examine the electric field carefully for perpendicularity in the vicinity of the conducting plane and then vary the position and magnitude of the image charge, i.e., mess up the solution, and notice the resulting departures from perfect orthogonality.

   The problem of a point charge in the presence of a conducting plane captures the essence of the method of images and these maneuvers can be generalized to deal with many other problems involving isolated charges and conducting surfaces. For example, Figure 4.3 shows a point charge a distance d from the center of a conducting spherical surface of radius $R(d > R)$. The method of images can be used to solve for the potential everywhere outside the surface in similar fashion.

   The solution to this problem, also given in most electricity and magnetism texts, involves placing a second charge of strength Q' (negative and different in magnitude than Q) somewhere inside the sphere—by symmetry one realizes that image charge must be placed somewhere along the line through the center of the sphere and Q—and showing that the resulting potential at distance R from the center of the sphere vanishes. (Or rather one determines the value of Q' and the distance, a of the charge from the center of the sphere such that Φ vanishes at $r = R$.) The precise solution will not be given here because this case is one of those available to solve by simulation in the IMAG&MUL program.

4.2.3   Extensions

The basic idea of the method of images was given in the preceding subsection. Once one has obtained the image solution to a particular problem, several other matters

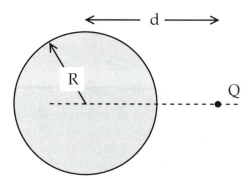

Figure 4.3: Charge $+Q$ located near a conducting sphere.

can be investigated. Let us return to the case of the point charge and conducting plane whose analytic solution was obtained through the method of images as Eq. 4.5. Since one has the potential for $x \geq 0$, one can deduce the electric field everywhere in that region using Eq. 4.3, or calculate it directly from the image solution using the standard expression for the electric field due to a point charge and the principle of superposition.

One might also be interested in the density of induced charge on the conducting surface. This is available using Gauss' law and the calculable electric field in the immediate vicinity of the conducting surface. For example, using Eqs. 4.3 and 4.5 we find the following expression for the field at $x = 0$:

$$\vec{E}(0, y, z) = -\frac{1}{4\pi\varepsilon_o}\left(\frac{2Qd}{(d^2 + y^2 + z^2)^{3/2}}\right)\hat{x}. \tag{4.6}$$

Gauss' law applied to a small "pillbox" shape protruding through the conducting interface then gives the induced surface charge density to be

$$\sigma(y, z) = \varepsilon_o E_x(0, y, z) = -\frac{1}{4\pi}\left(\frac{2Qd}{(d^2 + y^2 + z^2)^{3/2}}\right) \tag{4.7}$$

since the field vanishes inside the conductor.

Another application is the calculation of forces. In the context of the plane and charge problem, the (attractive) force on the charge is due to the negative surface charge density that developed on the conducting surface in response to the presence of the charge. That force can be determined by integrating the force exerted on the point charge by each small element of induced charge, or more cleverly in the characteristic image-method style. The force on the point charge is due to the part of the electric field not including its own contribution to the field. That field contribution must be exactly that of the *image charge!* Thus the force on the charge is trivially obtained from Coulomb's law to be

$$\vec{F} = -\frac{1}{4\pi\varepsilon_o}\frac{Q^2}{(2d)^2}\hat{x}. \tag{4.8}$$

It is a painful exercise in calculus to demonstrate that the same result is obtained from the integration of the $\hat{x}$ force components due to the surface charge density. It is worth emphasizing that this so-called *image force* is quite real. It must be taken into account, for example, in calculating the trajectories of charged particles moving in the vicinity of metallic surfaces.

It is also worth noting that the simulation is very useful for evaluating the image forces on charges. After one has found the solution to a particular situation it is easy using the program to take out the real charge. The E-field then presented by the program is that due to the image charge(s) and so when evaluated (using the *field probe*) at the (former) location of the real charge provides a determination of the force.

4.3   The Multipole Expansion in the Case of Azimuthal Symmetry

4.3.1   Introduction

Methods based on the multipole expansion allow solution of a far more general class of problems than the image charge approach. The calculations, however, are much more forbidding, except in situations of high symmetry. The simulation is restricted to problems of azimuthal symmetry, i.e., those in which the potentials, fields, and charge densities are independent of the angle ϕ of rotation about the z-axis. We have in mind the sort of situation displayed in Figure 4.4, i.e., a localized azimuthally symmetric equipotential surface. Our goal is to determine $\Phi(r, \theta)$ and the corresponding electric field outside the equipotential surface.

4.3.2   Basic Formulation

In spherical coordinates Laplace's equation (Eq. 4.1) takes the form

$$\nabla^2 \Phi = \frac{1}{r^2} \frac{\partial}{\partial r} \left(r^2 \frac{\partial \Phi}{\partial r} \right) + \frac{1}{r^2 \sin \theta} \frac{\partial}{\partial \theta} \left(\sin \theta \frac{\partial \Phi}{\partial \theta} \right) + \frac{1}{r^2 \sin^2 \theta} \frac{\partial^2 \Phi}{\partial^2 \phi} = 0 . \qquad (4.9)$$

Because we intend to treat only azimuthally symmetric situations we will eliminate the term arising from the ϕ variation. The remaining equation separates so that the solutions are product functions indexed by an integer l,

$$\Phi(r, \theta) = \sum_{l=0}^{\infty} \left(A_l r^l + \frac{B_l}{r^{l+1}} \right) P_l(\cos \theta) , \qquad (4.10)$$

where P_l is the lth Legendre polynomial. (See Table 4.1.)

We have in mind an *exterior problem*, i.e., a situation in which the potential is non-zero in some localized region and goes to 0 far away. If $\Phi(r \to \infty, \theta) \to 0$, then clearly we should let all the A_l's in Eq. 4.10 vanish, giving us the so-called multipole expansion

$$\Phi(r, \theta) = \sum_{l=0}^{\infty} \frac{B_l}{r^{l+1}} P_l(\cos \theta) , \qquad (4.11)$$

which will be the basis of what follows here and the multipole expansion (ME) mode of the simulation program.

First let us make clear the sense of that name. Consider the $l = 0$ term, B_0 / r. (The first several Legendre polynomials are given in Table 4.1; note that P_0 is just

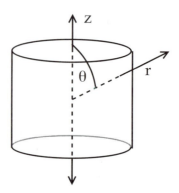

Figure 4.4: A cylindrical conducting surface held at potential Φ_o.

a constant.) Note that if we took $B_0 = Q/4\pi\varepsilon_o$ this term is precisely the potential due to a point charge Q located at the origin. Likewise, the $l = 1$ term, $B_1 \cos \theta/r^2$, is exactly the potential of a point dipole of dipole moment $\vec{p} = (4\pi\varepsilon_o)B_1 \hat{z}$ located at the origin, and the $l = 2$ term is the potential due to a point quadrupole, etc. We will discuss later what is meant, for example, by a *point* quadrupole; for the moment, note that we have observed that the potential outside the cylinder or any other azimuthally symmetric equipotential surface can be written as the sum of potential functions of multipole charge moments located at the origin. This does not mean, of course, that there really is any free charge at the origin; in the case at hand, it is actually distributed on the surface of a metallic object.

A (physical) dipole consists of equal and opposite charges, $\pm Q$, separated by a distance d. A *point dipole* is the limiting case of a physical dipole in which one imagines bringing the charges arbitrarly close to one another while increasing their strength such that the product Qd is constant. To see how this gives rise to the $l = 1$ term in Eq. 4.11 use Eq. 4.4 for the potential of a charge Q located at $(0, 0, z)$ and expand assuming $r \gg z$:

$$\Phi(r, \theta) = \frac{Q}{4\pi\varepsilon_o} \frac{1}{\sqrt{(r \sin \theta)^2 + (r \cos \theta - z)^2}}$$

$$= \frac{Q}{4\pi\varepsilon_o} \frac{1}{\sqrt{r^2 - 2rz \cos \theta + z^2}}$$

$$= \frac{Q}{4\pi\varepsilon_o r} \left(1 + \frac{z \cos \theta}{r} - \frac{1}{2}\frac{z^2}{r^2} + \frac{3}{2}\frac{z^2 \cos^2 \theta}{r^2} - \frac{3}{2}\frac{z^3 \cos \theta}{r^3} \right.$$

$$\left. + \frac{5}{2}\frac{z^3 \cos^3 \theta}{r^3} + \cdots \right) \tag{4.12}$$

where only terms of order z^3/r^3 have been kept. Adding the potentials in this form of charges $\pm Q$ located respectively at $z = \pm d/2$ gives

$$\Phi(r, \theta) = \frac{Qd}{4\pi\varepsilon_o r} \left(\frac{\cos \theta}{r} + \frac{d^2}{4} \left(\frac{5 \cos^3 \theta - 3 \cos \theta}{r^3} \right) + \cdots \right). \tag{4.13}$$

The $\cos \theta$ term, i.e., P_1, has emerged naturally and will survive the limiting process, $d \to 0$, $Q \to \infty$ such that $Qd \equiv p$ remains constant. (p is the dipole moment.)

Table 4.1: Legendre polynomials and pole nomenclature

l	$P_l(x)$	$n-pole$
0	1	*monopole*
1	x	*dipole*
2	$(3x^2-1)/2$	*quadrupole*
3	$(5x^3-3x)/2$	*octupole*
4	$(35x^4-30x^2+3)/8$	*hexadecapole*
5	$(63x^5-70x^3+15x)/8$	'32'-pole

Note also the octupole term ($l = 3$) lurking in Eq. 4.13. It disappears in the limit $d \rightarrow 0$ along with all higher-order terms. The above limit defines the *point dipole*. A *point quadrupole* is constructed similarly by combining equal and oppositely directed point dipoles separated by a short distance d and then allowing d to go to 0 with the product $pd \equiv \mathcal{Q}/4$ held constant and defining the quadrupole moment $\mathcal{Q}$. This process can be continued to visualize the higher-order point multipoles.

How does one use this approach, in particular Eq. 4.11, to solve for the potential given a situation like that displayed in Figure 4.4? If we can find the set of coefficients $\{B_l\}$ such that the potential is correctly given on the boundary, then the potential exterior to the boundary will be given uniquely by the sum in Eq. 4.11 with those values for the B_l's. Ordinarily (in textbooks, at least) one tries to determine the B_l's by some analytic means (see following subsection for an example), but it doesn't actually matter how you obtain them, of course. Getting them via computer simulation, as in the ME mode of the IMAG&MUL program, gives a perfectly valid solution and is a great intuition builder. The approach used will be as follows. Through the computer program you will control the values of the coefficients B_l while the computer shows a plot of the equipotential contours of $\Phi(r, \theta)$ computed from Eq. 4.11 superimposed on a picture of a selected equipotential surface (like the cylinder in Figure 4.4). Your object will be to adjust (by trial and error) the values of the B_l's to get the best agreement between the shape of that surface and a nearby equipotential contour. You will then have "solved" the problem, at least up to the limitations of the program, which incorporates only the first six l values. Having done so you will then have the full solution also displayed and can investigate its characteristics (like dependence on r, for example), determine the associated $\vec{E}$-field, etc.

In carrying out this adjustment process you will need to be able to visualize the angular dependence of the functions $P_l(\cos \theta)$. The polar plot of the P_l's in Figure 4.5 may be helpful. It shows polar plots of the angular dependence of the first six moments.

4.3.3   Example—The Split Sphere

To see multipole moments in action, let's consider an example that can be handled analytically and so is popular in textbooks. (See, for example, Jackson's graduate-level text,[6] which does this problem in several different ways.) Consider a conducting sphere which has been split in the *x-y* plane with an thin insulator

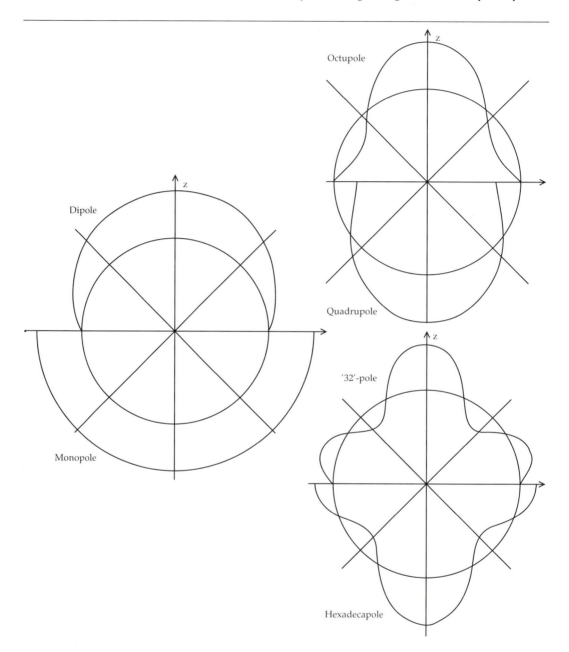

Figure 4.5: Polar plots of the first 6 P_l's. In each case half the function is shown; even (odd) l moments are even (odd) with respect to reflection in the x-y plane. Note that a polar plot does not show shapes or contours, but rather the values as a function of θ (measured from the z-axis plotted as a radius varying away from the unit circle, denoted by the solid circular curve in each panel); *negative values are plotted inside that unit circle!*

introduced between the hemispheres so that they can be held at differing potentials. Suppose further that the two hemispheres are being maintained at equal and opposite potentials, $\pm\Phi_o$. The problem is to determine Φ at all points in space

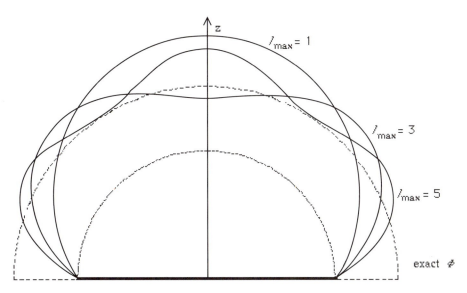

Figure 4.6: Polar plots of the convergence of Φ to the correct value of $\pm\Phi_o$ on the boundary of the split sphere.

exterior to the sphere. Clearly the potential will have somewhat of a dipole character since we have azimuthal symmetry and equal and opposite charge above and below the x-y plane. On the other hand, it cannot be just dipole in nature because the equipotential surfaces of the dipole potential are quite different from those the split sphere. Higher-order multipole moments will be needed to adjust the equipotential at the correct radius to be similar to that of the split sphere.

This situation has odd symmetry with respect to reflection in the x-y plane from which we realize that only the odd moments, dipole, octupole, "32"-pole, etc. will be needed. The correct potential must therefore consist of a sum as in Eq. 4.11 including only odd l terms. The specific values, deduced from an analytic approach that need not be described, here are given by the following beastly expression:

$$B_l = \left(-\frac{1}{2}\right)^{(l-1)/2} \frac{(2l+1)(l-2)!!}{2\left(\dfrac{l+1}{2}\right)!}, \qquad (4.14)$$

where the *double factorial* is defined via $(2n+1)!! = [(2n+1)(2n-1)\cdots 5\cdot 3\cdot 1$.

Eq. 4.14 can be evaluated to obtain $B_1 = 3/2$; $B_3 = -7/8$; $B_5 = 11/16$; etc. The convergence of the value of Φ to Φ_o on the boundary of the sphere is shown in Figure 4.6. Note that with increasing l_{max} the solution is fitting better the abrupt change in Φ at $\theta = \pi/2$. Clearly, however, one needs the higher moment contributions to sharpen the corner there. (This is analogous to the appearance of high Fourier components in the transform of a discontinuous function, like a square wave, for example.)

Only the upper half of the sphere is shown (the insulating layer is the bold line in the x-y plane); where the potential is positive. The inner dashed semicircle is the unit circle which denotes $\Phi = 0$ as in Figure 4.5. Note that all the approximations to the solutions vanish at the split.

4.3.4 Extensions

With mastery of the basic formalism of the multipole expansion and the simulation based approach for solving such problems, a variety of extensions are possible and interesting.

First, one can readily determine the associated electric fields, either by numerical gradient, or more accurately and efficiently by taking the analytic gradient of Eq. 4.11 term-by-term in order to obtain the multipole *field* expansion.

Another application of the multipole expansion is that much can be deduced about the charge distribution from the solution for the potential in a given case. Arbitrary localized charge distributions can be expanded in terms of their moments. For general charge distributions the monopole moment (the net charge) is a scalar, the dipole moment a vector, the quadrupole term a second-rank tensor, etc. Staying with azimuthal symmetry eliminates all but one term at each order, so that only a $\hat{z}$-component is possible for the dipole moment, only the Q_{33} of the quadrupole moment tensor survives, etc, and these quantities are given, in terms of the azimuthally symmetric charge distribution $\rho(r, \theta)$ as follows, for the first three moments:

$$Q = \int_{V'} \rho(\vec{r}')dv' \tag{4.15}$$

$$p(=p_z) = \int_{V'} z'\rho(\vec{r}')dv' \tag{4.16}$$

$$Q_{33} = \int_{V'} (3z'^2 - r'^2)\rho(\vec{r}')dv' \tag{4.17}$$

The potential expanded in terms of these moments is again just Eq. 4.11, expressed here in terms of the specific moments;

$$\Phi(r, \theta) = \frac{1}{4\pi\varepsilon_o}\left(\frac{Q}{r} + \frac{p}{r^2}P_1(\cos\theta) + \frac{1}{2}\frac{Q}{r^3}P_2(\cos\theta) + \cdots\right). \tag{4.18}$$

One way to exploit Eq. 4.18 then is to deduce something about the charge distribution giving rise to a given potential. Suppose for example that one has solved for the first six B_l's for the exterior potential of the cylinder in Figure 4.4. These known values then characterize the actual distribution of free charge. B_0 tells you the net free charge on the cylinder. B_1, by vanishing, indicates that the charge above the x-y plane balances that below in the sense of Eq. 4.16, etc.

4.4   Exercises

4.4.1 Image Charge Exercises

Most of these exercises are based on solutions of the specific image charge situations available in the simulation. (The situations available are described in section 4.3.) *Solution* in this context means the location and charge of all image charges used in a given situation.

4.1 **Warm-Up**

Use the simulation to verify the solution of the "point charge near conducting plane" example given in section 4.2.2 as follows. (The program defaults to this situation upon initiation.) Add a charge of strength -2 units at the appropriate position to 'solve' this problem.

    a. Using **Close-Up Window** tool, check the angle made by the field with respect to the conducting surfaces at five equally spaced distances from the line joining the charges to the edge of the screen. (If your solution is perfect the field should be everywhere perpendicular to the conducting surface. Make certain that you have placed the image charge with sufficient accuracy such that the angle with the surface is never less than 88 degrees.)

    b. Using the **Field Probe** tool, evaluate the field strength on the conducting surfaces at the following distances from the origin {0, 5, 10, 15, 20, 25}. N.B: The field probe also returns position information. Compare your results to Eq. 4.6.

4.2 **Cornered**

Consider a point charge located symmetrically in a square corner as in the case depicted as the "90-degree angle" situation in Figure 4.7.

    a. Use the simulation program to "solve" (in the sense defined above) this problem. Hint: You will need three image charges. Move the **Close-Up Window** tool to various points on the surface of the conductor to verify that the $\vec{E}$-field is everywhere normal to the conductor (within, say, 2 degrees).

    b. Remove the real charge from the situation (by just clicking on it). Now, as explained in section 4.2.3 , you have the field due just to the image charges, or equivalently to the surface charge induced on the surface of the conductor. Use the field probe to determine the magnitude and direction of this field. Compare your result to that obtained in the case of a charge in the vicinity of a single conducting plane using the same perpendicular distance from the planes. (N.B.: The simulation program uses different perpendicular distances for the two cases under consideration in the present problem.) Try to develop an explanation why the field strength is smaller for the corner than for the single plane.

    c. This situation can also be solved when the charge is not located symmetrically in the corner. Make a sketch showing where you think the image charges should be placed if the physical charge is located twice as far from one of the planes as the other. Check your answer using the simulation program; making a screen plot to show the $\vec{E}$-field for this case.

4.3 **Computing Forces I**

Find the solution of the 60 degree wedge problem (see Figure 4.8 later) using the simulation program. Based on the locations of the image charges in your

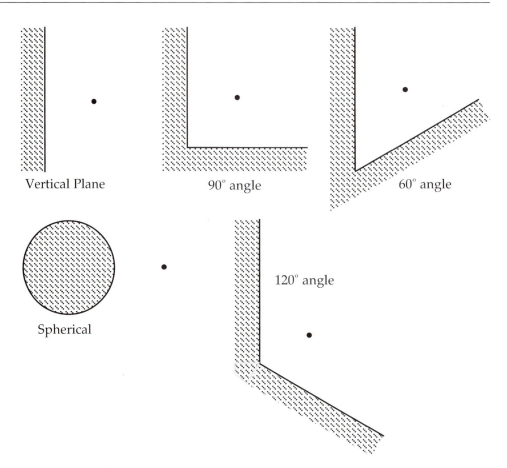

Figure 4.7: Available charge and conducting surface configurations for the IC mode of the simulation.

solution determine the magnitude and direction of the force on the real charge. (Take the magnitude of the real charge to be 10^{-6} C and its distance from the corner of the conductor to be 0.1 m.) Note that in the situation of a charge near a conductor that this force is actually exerted by the excess surface charge that accumulates on the surface of the conductor in response to the presence of the charge.

4.4   Point Charge and Conducting Sphere

Solve the "point charge near the conducting sphere problem" (see Figure 4.8) using the IMAG&MUL simulation program. (Hint: You'll need to vary both the strength and the position of the image charge to get an accurate solution.) Print out the resulting field distribution. From your solution can you also solve the corresponding problem of a point charge located inside a conducting spherical cavity? Describe the solution. Take and discuss the interesting limit of moving the point charge to the exact center of the cavity.

4.5   Computing Forces II

Using your solution of the sphere problem above determine the magnitude and direction of the force exerted on the real charge by the conducting sphere.

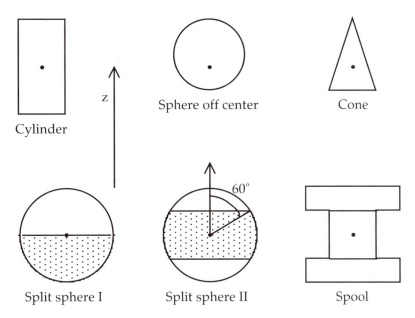

Figure 4.8: Available equipotential surfaces. Each is considered to be azimuthally symmetric. White areas are at potential Φ_o; shaded areas when present denote potential $-\Phi_o$.

Discuss the dependence of this force on the distance between the charge and the sphere. What answers do you expect for distances much more and much less than the sphere radius?

4.6 Computing Forces III, *or the Space Charge Sweeper*

(An analytic calculation thrown in for fun....) An electron is initially ($t = 0$) at rest a distance of 100 Å from a (perfectly conducting) metallic surface. A (pretend) image "positron" is therefore located 100 Å inside the surface and attracts the electron toward the metallic surface, causing it to enter and be absorbed by the conductor at time $t = t_o$. Determine that time t_o by integrating the equation of motion of the electron. (Answer: about 3×10^{-14} sec. This result explains why conductors are good at sweeping away space charge, i.e., stray electrons, in their vicinity while insulators are not.) Discuss the effect (qualitatively) of noninfinite conductivity in this problem. Hint: Think about Joule heating losses.

4.7 Induced Surface Charges I

The excess surface charge induced on the conducting surface by the nearby charge can be determined from your image charge solution and the known functional form of the field due to an isolated charge, as discussed in section 4.2.3.

    a. Use your image charge solution to the 60-degree wedge case, i.e., the positions and magnitudes of the image charges, to determine an analytic expression for the induced charge as a function of y, along the conducting surface that extends vertically from the origin. The excess surface charge induced on the conducting surface by the nearby charge can be determined numerically in the program, since it is proportional to the (normal component of the) field strength at the location of the conducting surface.

b. Using the **Field Probe** capability of the simulation program map out $\vec{E}$ along the conducting surface in the (solved) 60-degree wedge case. Print a copy of the screen. If you have done the preceding problem compare these numerical result (up to a constant of proportionality) to the analytic result.

4.8   Charge Between Parallel Planes

Suppose a point charge Q is located symmetrically between parallel conducting planes which are separated by distance d. The image charge method can be used to approximate the solution to this case as well.

a. Using the IMAG&MUL simulation program in the single-plane mode, move the real charge to a distance of x from the plane and imagine a second conducting plane drawn parallel to the first and at an equal distance to the right of the charge.

b. Approximate the solution to this problem by adding the four point charges outside of the region bounded by the planes that render the $\vec{E}$-field as perpendicular as possible to the two planes.

c. Using the field probe, determine the largest variation from orthogonality to the planes. How could you further diminish this error?

d. Again using the field probe to investigate the magnitude of the normal field component on the conductors, compare the induced surface charge density to that found for a single plane. Explain the differences you find between the single- and two-plane cases.

4.9   Image Dipoles

The image charge method can also be used to deal with the interaction of a dipole with a conducting surface. Using the IMAG&MUL simulation program in the single plane mode add a real charge equal and opposite in magnitude to the original real charge located a bit further from the plane, but along the same line perpendicular to it. Then add the appropriate image charges to solve for the field, and make a screen plot of your solution.

4.4.2   Multipole Expansion Exercises

The following exercises are based on solutions for particular situations set up in the multipole expansion mode of the IMAG&MUL simulation. For instructions on how to run the program and a list of available situations for solution see section 4.5.3.

4.10   The Cylinder

Solve the cylinder problem using the simulation; i.e., adjust the multipole moments to obtain the best (by eyeball*) match between the surface of the

---

*Don't spend hours refining your fit. The idea is to see how each of the six available multipole moments can contribute to improving the agreement.

cylinder and the nearest potential contour. (Hint: It works best to vary the lower-order multipole moments before the higher ones.) Print out a copy of the screen and include it with your solution.

  a. The limitation to $l = 6$ prevents a fully accurate solution. For what values of θ is the discrepancy largest and why?

  b. Use the field probe to determine the strength of the electric field along the z-axis and along the x-axis as a function of distance from the center of the cylinder. Make (superimposed) log-log plots of $\vec{E}(x)$ and $\vec{E}(z)$. The two curves tend toward the same power law dependence for large r. What is it and why?

  c. Estimate in terms of Φ_o the net charge on the cylinder.

  d. Switch to the **Display** option that shows the electric field. Where is the field the strongest?

4.11 The Split Sphere
Select the **Eq-pot Surfaces** option **Split Sphere I** and enter the known coefficient values given in section 4.3.3. Print out a copy of the screen.

  a. Use the field probe to determine the strength of the electric field along the z-axis as a function of distance from the center of the sphere. Make a log-log plot of the variation. What power law governs the dependence for large r in this case? Why?

  b. Estimate in terms of Φ_o the dipole moment of the sphere.

4.12 The Off-Center Sphere
Select the **Eq-pot Surfaces** option **Sphere Off Center** and adjust the moments to obtain a reasonably accurate solution. Print out a copy of the screen. Why is it necessary to include a dipole moment to match the equipotential accurately for this obviously symmetric object?

4.13 A Sphere With a Non-Zero Quadrupole Moment
Consider **Eq-pot Surfaces** option **Split Sphere II**. (This surface is shown in Figure 4.8. The angle shown there is exactly 60 degrees, and the shaded and unshaded parts of the surface are at equal and opposite potentials.)

  a. Formulate an argument for why the dipole moment must vanish for this equipotential surface.

  b. It is less straightforward, but also possible to show that the monopole moment vanishes for this case. If you are unable to establish this fact assume it and continue by adjusting in the program, the $l = 2 - 5$ moments for the best fit. Print the screen.

  c. Using the field probe determine and plot the variations of the field strength moving away from this surface in the x- and z-directions. What power law is in effect here?

4.14   Fields at Points on Conductors

Solve the cone case as well as you can using all six moments. Print the screen showing both the potential and the field. Using the field probe determine the magnitude of the field near the point of the cone, near the edge of the base of the cone, and near the center of the cone. If the potential Φ_o of the (conducting) cone is due to the presence of free charge, where is the concentration of charge the strongest? (This is the lightning-rod principle; see Ohanian's discussion.[5])

4.4.3   Possible Modifications to the Program

There are a number of extensions that can be made to this model to improve and broaden the physics content. The following subsections offer a partial list of possible extensions in the two program modes.

Image Charge Mode

4.15   Calculating forces

It is straightforward to add a calculation of the force on the real charge:

PROCEDURE Efield(Pt : RealPoint ; VAR EfieldX, EfieldY : real)

determines the field at position **Pt** by adding up the contributions of all the charges in a simple loop. Since the real charge is indexed as the first element of the arrays **ChargeOn**, **ChargeStrength**, and **ChargePts**, one can just start the incrementation at index 2 in order to leave out the field contribution of the real charge. From the field evaluated at the position of the real charge the force can be trivally calculated.

4.16   Line Charge Situations

The image charge method can also be used to treat some "line charge" situations, e.g., a line charge of charge per unit length λ located a distance d from a conducting plane. A little musing should convince you that the solution in this case is qualitatively similar to that of the point charge in the vicinity of a conducting plane; however, there is one important difference. The $\vec{E}$-field near a line charge has a different distance dependence than that near a point charge. This will change the appearance of the field lines as well as the image force and induced surface charge. It is rather straightforward to make the required program changes to deal with this variation by altering **PROCEDURE Efield** to reflect the correct distance dependence.

4.17   Induced Surface Charge Density

It would be elegant and aesthetic to add the capability of plotting the induced charge on all conducting surfaces. This can be determined as described in section 4.2.3 from the electric field evaluated at points on the conducting surface. See the CUPS utilities for information on plotting data.

4.18 More Than Two-Dimensional Corners

Although one could not easily display a three-dimensional (3-D) situation, one could rather readily adapt the simulation to deal with a charge in 3-D conducting corner. Charge locations then become 3-D variables that can be handled conveniently by altering variable type **RealPoint**. Distance calculations here and there would also need to be altered to take into account the third dimension. (From the perspective of physics, it is also fun to consider analytically the generalization to *N*-dimensional conducting corners.)

4.19 Dynamics

It would be more difficult but quite entertaining to allow the real charge — and its image(s)! — to move in response to the applied forces according to Newton's second law. One would add a new procedure, initiated by its own menu option which implements the following loop:

a. Determine the net force on the real charge and from that the vector acceleration. (The acceleration and velocity of the real charge should be set up as **RealPoint** type variables.) One needs, obviously, to assume a reasonable mass for the charge.

b. Adjust velocity of charge appropriately, based on a suitably chosen time step.

c. Adjust the position of the real charge according to the current velocity using the same time step.

d. Check whether the charge has moved into the conducting region. If it has, stop the calculation; if not, continue.

e. Move all image charges to the positions appropriate for that of the new charge. (This part would be specific to the situation chosen.)

f. Update the field display.

g. Test for satisfaction of an end condition.

Multipole Expansion Mode

4.20 Additional Surfaces

It is straightforward to program in additional azimuthally symmetric equipotential surfaces on which to perform the coefficient fits. One possibility, if you are working on this exercise in February is a heart shape, $\heartsuit$. (This will enable you to send the list of multipole moment coefficients to your physics friends instead of Valentine's Day cards*.)

4.21 Improving the Fit More Systematically

A more difficult extension would be to program a utility that determines the square deviation, χ^2, between the equipotential surface being fitted and the value of the actual potential integrated over the full surface.

---

*Due to the restriction to azimuthal symmetry, the shape in three dimensions is probably more reminiscent of a turnip than a heart, but perhaps the recipients will not think of that.

$$\chi^2 = \int_A (\Phi - \Phi_{fit})^2 \, dA \, . \tag{4.19}$$

This integral can be set up as a one-dimensional integral on the variable θ to be done numerically using any standard algorithm. (A truly elegant approach would be to use Gaussian quadrature[6] for the integration.) The value of χ^2 could then be used to improve the accuracy with which the B_l's are obtained by adjusting the latter to minimize the former.

4.22   "The Interior Problem"

Another extension would be to program instead the *interior* problem, i.e., the solution of Laplace's equation inside rather than outside a given boundary. In Eq. 4.10 one sets all the B_l's to zero and adjusts the values of the A_l's. **FUNCTION Potential** evaluates the sum in Eq. 4.10 and would require minor changes in the r-dependence. One would also, in implementing this extension, want to choose appropriate new potential surfaces and arrange to have them drawn in the main screen. This somewhat tedious task is carried out in **PROCEDURE DrawMESurface**.

4.5   Running the Program

4.5.1   General Introduction

The CUPS simulations are all written with a rather self-explanatory user interface and you will probably be able to explore the program without much assistance. The student is referred to Chapter 1 of this volume for general instructions on using the CUPS simulations and to the help screens available in the program for the particulars. (The help utilities are only available if the file IMAG&MUL.HLP is located in the directory or folder in which the simulation program resides. If the file is missing, check with your instructor.)

The simulation is written to run in two modes, image charge (IC) and multipole expansion (ME). It starts up by default in the IC mode. You can shift between the modes in the **PARTS** menu, available in either mode. Both modes include **FILE** and **PARTS** menu headings with the following uses.

- **FILE**: Controls the overall operation of the program in either simulation mode.

  - **About Program**: Displays the initial credit screen.

  - **About CUPS**: Displays information pertaining to the celebrated Consortium for Upper-Level Physics Simulations.

  - **Configuration**: Allows you to alter the configuration of your computer: colors may be substituted, the mouse "double-click" speed changed, the path for temporary files changed, etc.

  - **Exit Program**: Quits the simulation program.

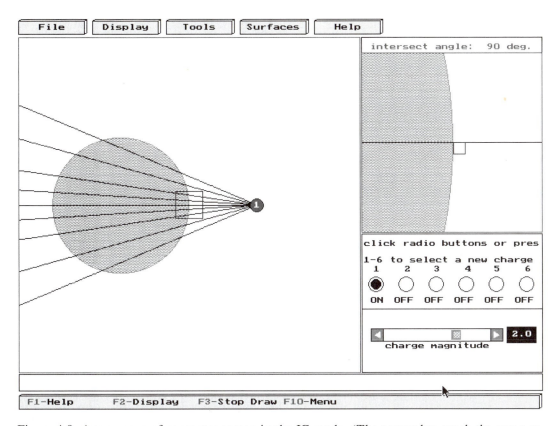

Figure 4.9: Appearance of computer screen in the IC mode. (The screen has much the same organization in the ME mode.)

- **PARTS**: Toggles between the available simulation modes.

  - **Multipole Expansion**: Selects ME mode.
  - **Image Charge**: Selects IC mode.

The appearance of the screen is similar in the two modes of operation and is shown in Figure 4.9. The main viewport shows the physical situation being considered. The menu bar is at the top and hot key definitions at the bottom. The panels on the upper and lower right are for display of detailed information and control of relevant variables, respectively.

The following subsections describe the menu options available in the two simulation modes.

4.5.2 The Image Charge (IC) Mode

The menu options available when the program is in the IC mode are as follows.

- **DISPLAY**: At all times the main display window shows the charge and conducting surface situation being investigated. The conducting region is always shaded in gray and the charge(s) (real and image) are denoted by small colored circles— red for positive, blue for negative. Physical charges must be located in the region not filled with the conducting medium; any charge shown in a conducting region is an *image charge*. This menu selection controls the display of the $\vec{E}$-field in that window. The field displayed is always that due to all charges, real and image, currently placed and corresponds to the correct electric field only if the conditions described in section 4.2.2 have been satisfied.

- **Show Charges Only**: Do not display the electric field at all.

- **Field Lines (in *vacuo* only)**: Displays electric field lines only in regions not filled with conducting medium. This is the physical field distribution since $\vec{E}$ must vanish within the conducting regions, although it is not the correct solution, of course, unless the proper image charges have been placed such that the field is everywhere perpendicular to any conducting surfaces.

- **Field Lines (all space)**: Displays the field due to all charges in all space. This is the $\vec{E}$-field one would have if the conducting regions were not actually present—i.e., that due to the real and image charges only.

- **TOOLS**: The menu controls the operation of other program utilities and display options for the other windows on the screen.

  - **Automatic Display**: An automatic display option (default is "on") is available under the **TOOLS** menu. Leaving this option on will result in redrawing of the fields and/or potentials in the main viewport whenever the situation has changed. If you prefer to control yourself when the main screen is updated you should turn the automatic display option off by simply clicking on that field under the **TOOLS** menu. When you wish to update the display, simply press hot key F2.

  - **Place Charges—Keyboard**: Opens an input screen for placing or altering charges.

  - **Move Close-Up Window**: The *close-up window* allows a magnified look at the $\vec{E}$-field (and conducting surface if one is within the close-up window) anywhere on the screen. The close-up tool is always active unless the field probe is in use and the magnified view is displayed in the upper right window. Choosing this selection allows one to move the close-up window anywhere within the main display simply by clicking and dragging or by using the **arrow** keys on your keyboard. The angle made by the electric field line with the conducting surface if the latter is within the close-up window is also determined and displayed above the magnified view.

  - **Field Probe**: The *field probe* is a tool which determines and displays in a separate window magnitude and direction of the $\vec{E}$-field at any particular point. Choosing this option activates the field probe. When active the field probe displays in the upper right-hand window the current location of the mouse and the field evaluated at that point. (Field is given in both Cartesian representation and in the form of magnitude and direction.) Clicking with the mouse at any desired location within the main window results in the display at that point of an arrow representing the magnitude and direction of the field.

- **Field Lines From Arb. Line**: The program ordinarily determines and plots only field lines originating any conducting surfaces present. In some cases the field therefore is not drawn in certain regions of the main screen. The **Field Lines from Arb. Line** utility allows the user to specify an imaginary line on the screen from which field lines are then drawn. The line is determined by clicking on any point, and the line defined is that one passing through the origin and the point clicked.

- **Refresh Screen**: Selection of this item causes the screen to be cleared and redrawn.

- **SURFACES**: This selection provides a menu of available conducting surface configurations. The available options are displayed in Figure 4.8.

  - **Vertical Plane**: This is the default configuration, a point charge in the presence of a conducting plane at $x = 0$.

  - **90-Degree Angle**: A point charge is located symmetrically in the vicinity of a 90-degree corner.

  - **60-Degree Angle**: A point charge is located symmetrically in the vicinity of an (acute) 60-degree corner.

  - **Spherical**: A point charge is located near a conducting spherical surface.

  - **120-Degree Angle**: A point charge is located symmetrically in the vicinity of a 120-degree corner.

- **HELP**: This selection provides help screens relating to other menu items.

4.5.3   The Multipole Expansion (ME) Mode

This section describes the menu options specific to the ME mode.

- **DISPLAY**: At all times the main window shows the "equipotential" surface being investigated, outlined in gray if the surface is to be assumed to be at the same (positive) potential everywhere. Two of the target surfaces have parts assumed to be at positive potential Φ_o and parts at an equal and opposite potential $-\Phi_o$. In those cases the part at $+\Phi_o$ is drawn with alternating gray and red, while that at $-\Phi_o$ uses gray and blue. (See Figure 4.9.)

  In addition the potential or the electric field or both are drawn as specified via this menu.

  - **Potential**: Displays the electric potential Φ in the form of a contour plot with contours equidistant in voltage plotted in color so that unconnected segments of the same contour can be readily identified. Red hues are used for more positive potentials, blue for negative.

- **Electric Field**: Displays $\vec{E}$-field lines.

- **Both Superimposed**: Displays both field and potential simultaneously.

• **TOOLS**: This menu controls the operation of other program utilities.

  - **Automatic Field Display**: See section 4.5.2.

  - **Change Moments—Keyboard**: Opens an input screen for altering multipole moments, B_l, using the keyboard. (Moments can also be altered with mouse using a screen with sliders.)

  - **Field Probe**: See section 5.8.1.

  - **Refresh Screen**: Clears and redraws the main window.

• **SURFACES**: This selection provides a menu of available target "equipotential" surfaces to which the contours are to be matched. All should be considered to be azimuthally symmetric with the cross section in the x-z plane being plotted in the program. The available surfaces in the program are displayed in Figure 4.9 and listed below.

  - **Cylinder**: (default)

  - **Split Sphere I**

  - **Sphere Off Center**

  - **Split Sphere II**

  - **Cone**

  - **Spool**

• **HELP**: This selection provides help screens relating to other menu items.

4.6   Program Structure

The simulation program accompanying this chapter has a somewhat complex organization owing to its multiple-part structure and the fact that the two program modes use some common utilities. The program procedures are grouped into the following sections, which are also clearly set off in the source code as well:

1. *Definition of TYPES, CONSTANTS, and GLOBAL VARIABLES*

2. *General utilities* used by both modes and in most cases having not too much to do with the physics of the simulations.

3. *IC mode "physics"*

4. *IC mode user interface* procedures. (Not much physics here; look at these only if familiar with the CUPS utilities.)

5. *ME mode "physics"*

6. *ME mode user interface* procedures. (Same *caveat* as above.)

7. *Main program* and initialization procedures. Again, not much physics.

Acknowledgments

Nathanial Johnson, Haverford College class of 1993, did much of the original programming for this simulation, contributed many insights, and taught me Pascal.

References

1. Griffiths, D. J. *Introduction to Electrodynamics*. 2nd ed. Englewood Cliffs, NJ: Prentice Hall, 1989.

2. Reitz, J. R., Milford, F. J., and Christy, R. W. *Foundations of Electromagnetic Theory*. 3rd ed. Reading, MA: Addison-Wesley, 1979.

3. Lorrain, P., Corson, D. R., Lorrain, F. *Electromagnetic Fields and Waves*. 3rd ed. New York: Freeman, 1988.

4. Jackson, J. D. *Classical Electrodynamics*. 5th ed. New York: Wiley, 1975.

5. Ohanian, H. C. *Physics*. 2nd ed. (expanded). New York: Norton, 1989. (Interlude VI.)

6. Press, W. H., Flannery, B. P., Teukolsky, S. A., and Vetterling, W. T. *Numerical Recipes: The Art of Scientific Computing*. Cambridge: Cambridge University Press, 1986.

5

Polarization and Dielectric Media

Lyle D. Roelofs
Physics Department, Haverford College

> We have seen that when electromotive force acts on a dielectric medium it produces in it a state which we have called electric polarization, and which we have described as consisting of electric displacement within the medium in a direction which, in isotropic media, coincides with that of the electromotive force, combined with a superficial charge on every element of volume into which we may suppose the dielectric divided, which is negative on the side toward which the force acts, and positive on the side from which it acts.

> —James Clerk Maxwell, 1891

5.1 Introduction

Maxwell's long sentence summarizes the response of nonconducting materials to electric fields, $\vec{E}(\vec{r})$. They become electrically *polarized* and as such give rise to additional contributions to the electric field. This conceptually simple idea acquires considerable complexity when fully implemented in that the problem must be treated self-consistently. To understand this in detail, consider a small unit of polarizable material, an atom or molecule, say,* within a larger sample. Take its location to be at the origin. If an electric field, $\vec{E}$, is applied to this element of material it develops a dipole moment,

$$\vec{p} = \alpha\vec{E}, \tag{5.1}$$

---

*Since polarization is a phenomenon of atomic origin, Maxwell's account, produced well before the advent of accurate models of the atom, is perhaps all the more remarkable.

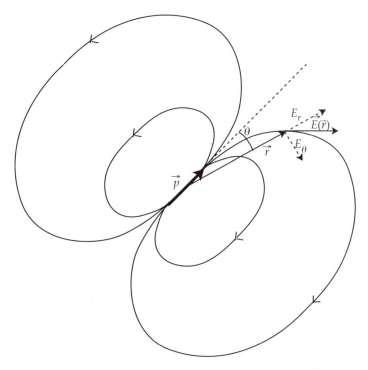

Figure 5.1: Field produced by a dipole moment $\vec{p}$.

parallel to the applied field. α is a constant for the atom or molecule in question and is called the "atomic polarizibility." Typical values for $\alpha/4\pi\epsilon_o$ range from 0.21×10^{-30} m^3 for helium, a noble gas element, to 34×10^{-30} m^3 for potassium, from the readily polarized alkali metal column. (The origin of this polarization is discussed in the next section.) The induced dipole moment results in a new contribution to the electric field throughout the sample. The field can be deduced from the expression for the potential of a point dipole at the origin:

$$\Phi(\vec{r}) = \frac{\vec{p} \cdot \vec{r}}{4\pi\epsilon_o \, r^3}. \tag{5.2}$$

The associated electric field is obtained by taking the gradient

$$E_r = \frac{2p \cos \theta}{4\pi\epsilon_o \, r^3} \tag{5.3}$$

$$E_\theta = \frac{p \sin \theta}{4\pi\epsilon_o \, r^3}. \tag{5.4}$$

(In Eqs. 5.3 and 5.4, θ represents the polar angle relative to the direction of the dipole moment, $\vec{p}$, as illustrated in Fig. 5.1) Thus the rest of the sample experiences an altered electric field due to the polarized atom. If there is polarizable material at other locations, its polarization is altered by this change in the field, and that in turn changes the applied field at initial location, and so on. The simulations associated with this chapter explore this subject at two levels: ATOMPOL, at the atomic or molecular scale where the phenomenon arises; and DIELECT, at the macroscopic scale where the phenomenon is described via the polarization field, $\vec{P}(\vec{r})$, and the electric displacement, $\vec{D}(\vec{r})$. Background material for these simulations is covered here; for further information and detail consult any of the standard text-books in this area, Griffiths,[1] Reitz, Milford, and Christy,[2] and Lorain, Corson, and Lorain,[3] among others.

5.2   *The Atomic Origin of Polarization*

Polarizibility arises both at the atomic and molecular levels. Some molecules— H_2O is the typically noted example—exhibit a permanent electric dipole moment due to charge transfer accompanying the chemical bonding of the molecule. The torque induced by an applied electric field on such a molecule can be expressed in terms of its dipole moment:

$$\vec{\tau} = \vec{p} \times \vec{E}. \tag{5.5}$$

The molecular moments then tend* to align with the field, thus leading to polarization of the medium containing them. Depending of course on the magnitude of the permanent dipole moment, this sort of response can be quite strong and materials made up of these so-called polar molecules are among the most polarizable sub-stances found in nature. Note that because this response depends on the molecules being able to rotate freely, the liquid phase of such materials is usually the most readily polarized. This mode of polarization is responsible for several very im-portant effects. The ability of water to dissolve salts is due to the attractive inter-action that occurs between the separated ions and the appropriately oriented water molecules. Another application is the formation of cell walls and other thin mem-branes made up of long molecules that obtain their layer-forming properties by be-ing *polar* (thus being drawn toward water) at one end and *oily* (thus avoiding contact with water) at the other. (For an introductory-level treatment of the formation of such layers see Styer,[4] for example.) Despite the interest of this sort of polarizabil-ity, its complexity (position and orientation of the molecules as well as their elec-tric interactions must be faithfully treated) rules out its treatment except by much more powerful computers than those this project is designed for.

There is another origin of polar behavior, one that occurs in all atoms and molecules.† The application of an electric field causes the electrons and nuclei to experience forces in opposite directions and thus to displace oppositely. The dis-placements, though small, produce a dipole moment that results in a non-negligible field contribution. The physics of this process is aptly illustrated by considering as a model of an atom, a point nucleus of charge $+Ze$ surrounded by a sphere of uniformly distributed negative charge density representing the electrons. Take the sphere to be of radius R. It is left as an exercise to show that if the nucleus is dis-placed by amount d from the center of the electron cloud ($d < R$), a restoring force of magnitude

$$F_{res} = \frac{Z^2 e^2 d}{4\pi\epsilon_o R^3} \tag{5.6}$$

---

*Thermal agitation prevents perfect alignment.

†It is masked by the stronger response in those materials whose molecules exhibit permanent moments.

acts to restore the nucleus to its rightful position. Under the application of an external field the equilibrium position of the nucleus, relative to the center of the charge cloud, is the position where the force due to that external field is balanced by the force induced by the polarization of the atom:

$$d_{eq} = \frac{4\pi\epsilon_o R^3}{Ze} E_{ext}. \tag{5.7}$$

The induced dipole moment of the atom is the charge times the separation:

$$p = d(Ze) = 4\pi\epsilon_o R^3 E_{ext}. \tag{5.8}$$

Comparing Eqs. 5.1 and 5.8 gives the polarizability of our model "atom":

$$\alpha = 3v\epsilon_o, \tag{5.9}$$

where v is the volume of the atom. Note that

$$\frac{\alpha}{4\pi\epsilon_o} = R^3, \tag{5.10}$$

so that it is convenient to specify that quantity in units of $\mathring{A}^3$ or 10^{-30} m^3 as was done in section 5.1.

5.2.1   Interactions

The interaction between polarized atoms or molecules via the electric field they generate are explored in the ATOMPOL simulation. Some of the key features to be expected are discussed in this section.

First, note that since the dipole field decays like $1/d^3$, the interactions are relatively short in range. Secondly, note that there is a strong directional anisotropy. Let $\vec{r}$ denote the separation between two polarizable atoms. Because of the nature of the field associated with a dipole (see Fig. 5.1) the total field experienced by each dipole is enhanced if $\vec{r}$ is parallel to $\vec{E}$ and reduced if the two vectors are perpendicular to one another. One might then wonder which effect is more important in a material made up of a uniform distribution of such moments. We can answer this question by averaging the component of the dipole field Eqs. 5.3 and 5.4, parallel to $\vec{p}$ itself. Let the moment be along the z-direction. Then

$$E_z(r, \theta) = \frac{p(3\cos^2\theta - 1)}{4\pi\epsilon_o r^3}. \tag{5.11}$$

If we average this field component by integrating over all space outside a sphere of radius R centered on the dipole moment, we find that the integral vanishes. The answer to our question is then as follows. A polarized atom tends to increase the field in regions "ahead and behind" it and to decrease the field in the sideways direction, but these two effects cancel when averaged over all directions.

Despite this overall averaging away, there are pronounced local effects that will be investigated in the simulation and one important global effect missed by the above construction which will be discussed in section 5.5 , where it will be seen that in most cases, polarization results in an overall reduction of the electric field strength in the material.

5.2.2 Iterative Solution

The interactions described in the previous section are a complication to determining the total electric field when several or many polarizable atoms are present in close proximity to one another. We cannot determine the total field unless we know all the polarizations and we cannot determine the polarizations until we know the electric field. Fortunately this problem can be approached iteratively as described in this section.

A step-by-step approach that converges to the solution of the combined field/polarization problem works as follows. Let a group of atoms be immersed in an externally applied electric field, $\vec{E}_{ext}$, and let $\vec{p}_i$ denote the polarizability of the ith atom.

1. Begin with all $\vec{p}_i$'s set equal to $\alpha \vec{E}_{ext}$. Then loop through all the atoms performing the following steps until no further changes occur.

2. At atom i determine the total electric field consisting of the external field added to the field contributions due to the polarization of the other atoms. These field contributions will depend on the relative positions of the atoms via Eqs. 5.3 and 5.4.

3. Using this value of the electric field and Eq. 5.1, update $\vec{p}_i$.

4. Test for convergence by checking the difference between the previous and new polarizations of each atom. The solution is considered to be complete when in any full sweep through all cells, no polarization has changed by more than a preset tolerance. Otherwise loop to step 2 and continue.

The program ATOMPOL implements this iterative approach allowing investigation of many aspects of polarization and the interaction of polarized entities.

5.2.3 Preventing Polarization Divergences

As developed in exercise in this chapter, Eq. 5.1 can result in divergent increases in polarization if sufficiently polarizable atoms are located in close proximity. In the ATOMPOL program the possibility of a divergent polarization is finessed via an enforced saturation of the field acting at each atom location. Specifically, the update of the polarization is done using

$$p_x^{new} = E_{max}\alpha \tanh\left(E_x/E_{max}\right) \tag{5.12}$$

$$p_y^{new} = E_{max}\alpha \tanh\left(E_y/E_{max}\right). \tag{5.13}$$

Note that if the field components are considerably less than the parameter E_{max}, these reduce to Eq. 5.1, while if the fields components get large, the polarization is limited to αE_{max}. The value of E_{max} is set to 100 in the program. (See section 5.8.2 for a discussion of the units used in the program.)

5.3 Macroscopic Polarization

5.3.1 Polarization Field and Susceptibility

Next we consider this phenomenon on a larger scale, defining a volume element ΔV to be our basic unit of consideration and taking it to be of sufficient size such that it contains many individual (polarizable) atoms or molecules. We characterize the degree of polarization of the material by defining the *polarization field* to be

$$\vec{P} = \frac{\Sigma \vec{p}}{\Delta V}, \tag{5.14}$$

i.e., the total dipole moment per unit volume. Note that the units of $\vec{P}$ are $[charge]/[length]^2$, which are the same as the units of $\epsilon_o \vec{E}$.*

If $\vec{p}$ is linearly related to $\vec{E}$ as in Eq. 5.1, one might expect that $\vec{P}$ is also proportional to $\vec{E}$, and indeed it is for many materials. The materials are known as *linear dielectrics* and the constant characterizing the linear relation is termed the *susceptibility* χ:

$$\vec{P} = \epsilon_o \chi \vec{E}. \tag{5.15}$$

(Students using Reitz, Milford, and Christy[2] should beware of the use of a nonstandard definition of susceptibility in that text. Identical definitions of $\vec{E}$ and $\vec{P}$ are used, but in the equation relating them the factor ϵ_o is suppressed. Thus their χ is not dimensionless, but instead includes a factor of ϵ_o.)

To understand macroscopic polarization, we need to determine the relation between χ and the atomic polarizibility, α, associated with the polarizable units of which the material consists.

5.3.2 Connection Between Atomic and Macroscopic Polarization

One might first guess on the basis of Eqs. 5.1, 5.14, and 5.15 that α and χ are simply related through the atomic density, n, the number of atoms or molecules per unit volume in the material. If that were the case we would have

$$\vec{P} = n\vec{p} \Longrightarrow \epsilon_o \chi \vec{E} = n\alpha \vec{E} \Longrightarrow \chi = \frac{n\alpha}{\epsilon_o}. \tag{5.16}$$

This is incorrect in that we need to be careful about accounting for the fields due to individual dipoles and not allow them to "act on themselves." The $\vec{E}$ which appears in Eq. 5.15 is the actual electric field in the material including the contributions of all polarized entities. The electric field in Eq. 5.1 defining atomic polarizability, however, should not include the field of the dipole itself. Thus we could rewrite that equation in terms of the same electric field as used for Eq. 5.15 as

$$\vec{p} = \alpha(\vec{E} - \vec{E}_{self}) \tag{5.17}$$

---

*For this reason the simulation calculates and plots $\vec{P}/\epsilon_o$ rather than $\vec{P}$.

where $\vec{E}_{self}$ is the field due to the dipole itself and is therefore proportional to $\vec{p}$. This self-field can be determined to be that of the negative charge cloud evaluated at the location of the nucleus. We find (see Eq. 5.1) that it can be written

$$\vec{E}_{self} = -\frac{\vec{p}}{4\pi\epsilon_o R^3}, \tag{5.18}$$

where R is again the atomic radius. We substitute Eq. 5.18 into Eq. 5.17 to obtain

$$\vec{p} = \alpha\left(\vec{E} + \frac{\vec{p}}{4\pi\epsilon_o R^3}\right), \tag{5.19}$$

so that

$$\vec{p} = \left(\frac{\alpha}{1 - \alpha/4\pi\epsilon_o R^3}\right)\vec{E}. \tag{5.20}$$

Now that $\vec{E}$ means the same thing, Eqs. 5.15 and 5.20 imply that

$$\chi = \frac{n\alpha/\epsilon_o}{1 - \alpha/4\pi\epsilon_o R^3}. \tag{5.21}$$

If the atoms fill the space their density is simply related to the volume per atom*, $n = 1/v$, so that

$$\frac{1}{4\pi R^3} = \frac{n}{3}, \tag{5.22}$$

leads finally to

$$\chi = \frac{n\alpha/\epsilon_o}{1 - n\alpha/(3\epsilon_o)}. \tag{5.23}$$

Approximations have been made in this derivation so that Eq. 5.23 should be regarded as an instructive estimation, not an accurate prediction.

5.3.3   A Cell-Based Approach

Equation 5.14 suggests a cell-based approach to the phenomenon of macroscopic polarization. Let us partition space into cells of dimension $a \times a \times a$, choosing a sufficiently small such that within each cell the electric field, $\vec{E}$, and polarization, $\vec{P}$, are approximately uniform, but large enough so that each cell still contains many atomic dipoles. (If the cell contains no polarizable material, $\vec{P} = 0$, of course.) We then identify the ΔV of Eq. 5.14 to be a^3. An approach based on dividing up the material into cells has the additional advantage that it finesses the very complicated field distribution at the atomic scale, replacing it by a smooth average.

A cell-based approach is natural for simulating dielectric behavior in that it provides a spatial lattice on which to define the fields and a framework for iterating the solution to self-consistency. Let i, j, and k be the cell indices in the x-, y-, and z-directions, respectively, and let $\vec{E}^{(i,j,k)}$ and $\vec{P}^{(i,j,k)}$ represent the electric field

---

*One needs a correction factor if the atoms do not pack tightly in the solid.

and polarization in cell (i, j, k). Define also array $\chi^{(i,j,k)}$ to represent the susceptibility of the material in cell (i, j, k), taking $\chi^{(i,j,k)} = 0$ in any empty cells. A self-consistent solution to a dielectric problem involves determining $\vec{E}^{(i,j,k)}$ and $\vec{P}^{(i,j,k)}$ in all cells such that Eq. 5.15 holds in each cell. Note that $\vec{E}^{(i,j,k)}$ must include the contributions of external fields, the fields due to the non-zero polarization of other cells and the field due to the polarization of cell (i, j, k) itself, i.e.,

$$\vec{E}^{(i,j,k)} = \vec{E}_{ext} + \vec{E}^{(i,j,k)}_{othercells} + \vec{E}^{(i,j,k)}_{pol}, \tag{5.24}$$

respectively.

An iterative scheme that will converge to this self-consistent solution is as follows.

1. Begin with all $\vec{P}^{(i,j,k)} = 0$ and all $\vec{E}^{(i,j,k)}$ set equal to $\vec{E}_{ext}$, the applied external field. Then loop through all the cells with $\chi^{(i,j,k)} \neq 0$, performing the following steps until no further changes occur.

2. At cell (i, j, k) determine the field $\vec{E}^{(i,j,k)}_{othercells}$ in the cell due to the polarization of all the other cells by summing up their field contributions using Eqs. 5.3 and 5.4.

3. Based on the value of $\vec{E}^{(i,j,k)}_{othercells}$ update $\vec{P}^{(i,j,k)}$*

4. Test for convergence by checking the difference between the previous and new polarization values in the site. The solution is considered to be complete when in any full sweep through all cells, no polarization has changed by more than a preset tolerance. Otherwise loop to step 2 and continue.

Step 2 in the above iteration requires a calculation of the cell polarization in response to the applied field. Eq. 5.15 is not sufficient for this purpose, since the $\vec{E}$ referred to in that equation includes also the field induced by the polarization of the cell, $\vec{E}_{pol}$. We consider this field next.

5.3.4   Field Within a Polarized Cell

Within a polarized cell there are two contributions to the total electric field: $\vec{E}_{outside}$, i.e., the field due to all external sources, and $\vec{E}_{pol}$, that due to the polarization of the cell itself. For the purposes of calculating and plotting the total field one needs to characterize the latter in terms of $\vec{P}$.

Consider Fig. 5.2, which shows a cell and the $\vec{P}$ in it. A straightforward argument from vector calculus[1] shows that the potential due to the polarization within a cell is identical to that produced by a surface charge density distribution

$$\sigma_b = \vec{P} \cdot \hat{n} \tag{5.25}$$

located on the surfaces of the cell. ($\hat{n}$ is the usual outward-directed unit vector for each surface of the cell.) The subscript b on the charge density in Eq. 5.25 denotes "bound" since this induced charge density is not free to move about the

---

*Changing $\vec{P}^{(i,j,k)}$ changes the field applied in all other cells. Hence the necessity for several (many?) passes through the entire lattice.

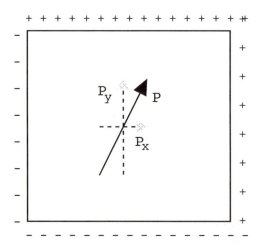

Figure 5.2: A cell of uniform polarization showing the $\vec{P}$ vector, its components and the equivalent polarization surface charge density on the boundaries of the cell.

material as are, for example, the conduction electrons in a metal. Rather it represents a local non-zero charge density due to the displacement in different directions of positive and negative charge that remains associated with the individual atoms or molecules comprising the material.

The average electric field within the cell due to the polarization, $\vec{E}_{pol,}$ can then be deduced from the equivalent bound charge distribution in Eq. 5.25. Thus one regards the boundaries of the cell in Figure 5.2 as being small capacitor plates bounding the cell and giving electric field components in each direction. Expressing this field in terms of $\vec{P}$ gives

$$\vec{E}_{pol} = -\beta \frac{\vec{P}}{\epsilon_o}, \tag{5.26}$$

where β is a factor that accounts for the finite dimensions of the cell. $\beta \cong 0.337$. In section 5.2.1 it was noted that polarization of a material usually results in a reduction of the electric field within the material if the external charge density to which the field is due is not changed.* The reduction is due to the negative field contribution found in Eq. 5.26.

5.3.5   Cell Polarization Response to an "Outside" Field

In order to implement the iterative scheme of section 5.3.3 one needs to determine $\vec{P}^{(i,j,k)}$ not from the total field within the cell as in Eq. 5.15, but from $\vec{E}_{outside}$. This can be done as follows. Recall that on the basis of the definitions of the different field contributions

$$\vec{E} = \vec{E}_{pol} + \vec{E}_{outside}. \tag{5.27}$$

(Cell indices will be suppressed in this section as all refer to the same cell.) Use Eq. 5.26 to replace $\vec{E}_{pol}$:

---

*Hence the practice of filling capacitors with polarizable materials. The field is reduced thereby decreasing the potential difference between the plates and thus increasing the capacitance $C = Q/V$. Inclusion of material between the conducting plates also allows for convenient control of the plate separation.

$$\vec{E} = -\beta \frac{\vec{P}}{\epsilon_o} + \vec{E}_{outside}. \qquad (5.28)$$

Next we use Eq. 5.15 to replace $\vec{P}$ to find

$$\vec{E} = -\beta \chi \vec{E} + \vec{E}_{outside}, \qquad (5.29)$$

and simplify to obtain

$$\vec{E} = \frac{\vec{E}_{outside}}{1 + \beta \chi}. \qquad (5.30)$$

Hence, using Eq. 5.15 again, $\vec{P}$ can be expressed in terms of $\vec{E}_{outside}$ as

$$\frac{\vec{P}}{\epsilon_o} = \frac{\chi}{1 + \beta \chi} \vec{E}_{outside}, \qquad (5.31)$$

so that a cell polarization can be obtained in terms just of the field due to external sources and the other cells. Eq. 5.31 implements step 2 in the self-consistency iteration of section 5.3.3.

5.3.6   Adjacent Polarized Cells and $\vec{D}$

Basic texts in electricity and magnetism also define a third field, $\vec{D}$, the *electric displacement*. The DIELECT simulation also determines $\vec{D}$. The definition of $\vec{D}$ and its usefulness are considered in this section.

We begin by considering what happens when two polarized cells are placed end to end. Suppose only cells (i, j, k) and $(i + 1, j, k)$ are polarized (equally) and in the x-direction. (Take the cell index i to run in the horizontal direction across the screen, the x-direction, and j to be the vertical cell index, i.e., the y-direction.) Let the polarizations be respectively, $P_x^{(i,j,k)}$ and $P_x^{(i+1,j,k)}$. We will take Eqs. 5.3 and 5.4 to give the field due to a particular polarized cell.*

To determine the field due to two adjacent cells note that by Eq. 5.25 the cell polarizations give rise to bound charge densities, ρ_b, of equal magnitude, with signs as shown in Figure 5.2. Note that at the shared face of the two cells the net bound charge density is then $P_x^{(i,j,k)} - P_x^{(i+1,j,k)}$, so that one has perfect cancellation if the polarizations are equal. In that case the field of the two-cell combination would be that of the two exterior bound charge planes separated by two cell lengths—so that the dipole moment is twice that of either individual cell—as if the shared surface were not present at all.

If $P_x^{(i,j,k)} - P_x^{(i+1,j,k)} \neq 0$, then there is net bound charge of that magnitude at the interface between the cells. Note that the "discrete divergence"† of $\vec{P}$ in this geometry would be

$$\vec{\nabla} \cdot \vec{P} \equiv \frac{P_x^{(i+1,j,k)} - P_x^{(i,j,k)}}{a} = -\frac{\sigma_b^{(net)}}{a} = -\rho_b, \qquad (5.32)$$

---

*Note that we are thereby treating the field of each cell as a point dipole field. This approximation is accurate for regions relatively far from the cell producing the field, but is not realistic nearby due to the fact that the bound charge producing the moment is should be considered to be spread over the surfaces of the cell, rather than concentrated at its center.

†That is, the divergence approximated on our grid using a finite difference.

where a is the cell dimension and ρ_b thus represents the 3-D density of bound charge. Eq. 5.32 resembles Gauss' law for the electric field,

$$\vec{\nabla} \cdot \vec{E} = \frac{\rho}{\epsilon_o}, \tag{5.33}$$

and can be rewritten

$$\vec{\nabla} \cdot \frac{\vec{P}}{\epsilon_o} = -\frac{\rho_b}{\epsilon_o} \tag{5.34}$$

Note that Eqs. 5.34 and 5.35 can be combined in an intriguing fashion

$$\vec{\nabla} \cdot (\epsilon_o \vec{E} + \vec{P}) = \rho - \rho_b, \tag{5.35}$$

or

$$\vec{\nabla} \cdot \vec{D} = \rho_f, \tag{5.36}$$

where $\vec{D}$ is the electric displacement defined via

$$\vec{D} = \epsilon_o \vec{E} + \vec{P}. \tag{5.37}$$

and ρ_f is the density of free charge, i.e., the charge density not including any polarization-induced bound charge. The utility of $\vec{D}$ is that the Eq. 5.36 is Gauss' law in terms only of the free charge density. In some symmetrical situations one can thus solve for $\vec{D}$ more readily than for either $\vec{P}$ or $\vec{E}$. The DIELECT simulation can be used to verify Eq. 5.36 for arbitrary situations involving polarized material.

5.4   A Two-Dimensional Implementation

To allow convenient visualization and more rapid computation, the dielectric systems simulated in DIELECT are restricted to being uniform in the z-direction. This ensures that all fields and polarizations are vectors in the x-y plane, which in the display is taken to be the plane of the screen. Therefore each displayed cell in the screen is actually one of an infinite *"z-column"** of cells extending perpendicular to the screen.

One must therefore keep in mind that the actual field produced by one displayed polarized cell is really that of a full z-column of cells of identical polarization. A second complication is that when we adjust the polarization of a cell, we are really adjusting the polarization of whole z-column of cells and we need therefore to take into account the effect of the electric field of the other cells in the z-column. This subtly alters the update Eq. 5.31 in section 5.3.5. These two complications are sorted out in this section. z-Columns can be indexed with just their x, y-coordinates, i.e., (i, j). The notation used in this section and following is that any vector with a two-variable index denotes the value of that quantity in an entire z-column.

---

*We use this terminology to distinguish such a column from a column of cells in the vertical direction in the plane of the screen, i.e., the y-direction.

5.4.1 The Field of a *z*-Column

It is convenient to work in a Cartesian representation so that the field due to a dipole moment, given in spherical coordinates in Eqs. 5.3 and 5.4, should be re-written. It is left as an exercise to show that if the dipole components are p_x and p_y, the field contribution at position (x, y) of a dipole, $\vec{p} = (p_x, p_y, 0)$, at the origin is

$$E_x(x, y) = \frac{(2x^2 - (y^2 + z^2))p_x + 3xyp_y}{4\pi\epsilon_o r^5}$$

$$E_y(x, y) = \frac{3xyp_x + (2y^2 - (x^2 + z^2))p_y}{4\pi\epsilon_o r^5} \qquad (5.38)$$

$$E_z(x, y) = \frac{3xzp_x + 3yzp_y}{4\pi\epsilon_o r^5},$$

where r, as in Eq. 5.2, denotes the distance from the dipole,

$$r = (x^2 + y^2 + z^2)^{1/2}. \qquad (5.39)$$

Using Eq. 5.38 we wish to determine the electric field at z-column (i, j) due to all the cells in z-column (i', j'). Averaging over all the dipoles in cell (i', j', k') allows us to express the field in terms of $\vec{P}^{(i',j',k')}$ using Eq. 5.14:

$$\sum_{cell\ (i',j',k')} p_x = a^3 P_x^{(i',j',k')}. \qquad (5.40)$$

Eq. 5.38 puts cell (i', j', k') at the origin, so

$$x = (i - i')a \equiv \Delta_x,$$

$$y = (j - j')a \equiv \Delta_y, \qquad (5.41)$$

$$z = (k - k')a.$$

Written fully in terms of indices, the field at any cell in the (i, j)th z-column due to $\vec{P}(i', j', k')$ is

$$E_x^{(i,j)}(i', j', k') = \frac{(2\Delta_x^2 - (\Delta_y^2 + k'^2))P_x^{(i',j',k')} + 3\Delta_x\Delta_y P_y^{(i',j',k')}}{4\pi\epsilon_o(\Delta_x^2 + \Delta_y^2 + k'^2)^{5/2}} \qquad (5.42)$$

$$E_y^{(i,j)}(i', j', k') = \frac{3\Delta_x\Delta_y P_x^{(i',j',k')} + (2\Delta_y^2 - (\Delta_x^2 + k'^2))P_y^{(i',j',k')}}{4\pi\epsilon_o(\Delta_x^2 + \Delta_y^2 + k'^2)^{5/2}}. \qquad (5.43)$$

Note that since $\vec{E}^{(i,j,k)}$ cannot depend on k, we have taken $k = 0$ in the above and suppressed that index. Moreover, the z-component of $\vec{E}^{(i,j,k)}$ must cancel after summing on k', so we have not bothered to write the contribution $E_z^{(i,j)}$.

It remains to sum over k' to determine the entire effect of z-column (i', j') on any cell in z-column (i, j):

$$E_x^{(i,j)}(i', j') = \sum_{k'=-\infty}^{\infty} E_x^{(i,j)}(i', j', k')$$

$$= \frac{1}{4\pi}\left(\frac{P_x^{(i',j')}}{\epsilon_o} S_1(\Delta_x, \Delta_y) + \frac{P_y^{(i',j')}}{\epsilon_o} S_2(\Delta_x, \Delta_y)\right), \qquad (5.44)$$

where S_1 and S_2 represent the sums

$$S_1(\Delta_x, \Delta_y) = \sum_{k'=-\infty}^{\infty} \frac{2\Delta_x^2 - (\Delta_y^2 + k'^2)}{(\Delta_x^2 + \Delta_y^2 + k'^2)^{5/2}} \tag{5.45}$$

$$S_2(\Delta_x, \Delta_y) = \sum_{k'=-\infty}^{\infty} \frac{3\Delta_x \Delta_y}{(\Delta_x^2 + \Delta_y^2 + k'^2)^{5/2}}. \tag{5.46}$$

These sums are unpleasant, but can be evaluated numerically.* The needed values are given in a data file, DIELEC.DAT, provided with the simulation programs. This file must be present for the DIELECT program to run. The y-component of the field can also be expressed in terms of S_1 and S_2:

$$E_y^{(i,j)}(i', j') = \frac{1}{4\pi}\left(\frac{\vec{P}_y^{(i',j')}}{\epsilon_o} S_1(\Delta_y, \Delta_x) + \frac{\vec{P}_x^{(i',j')}}{\epsilon_o} S_2(\Delta_y, \Delta_x)\right). \tag{5.47}$$

N.B: $S_2(\Delta_y, \Delta_x) = S_2(\Delta_x, \Delta_y)$, but the same does not hold for S_1.

5.4.2   The "Self-Field" of a z-Column

We previously defined $\vec{E}_{outside}^{(i,j)}$ to be (see section 5.3.4) the field in cell (i, j) due to all sources not including the polarization of that cell. Let us now further resolve $\vec{E}_{outside}^{(i,j)}$ into the following contributions: from external sources, $\vec{E}_{ext}$; from other z-columns of cells, $\vec{E}_{other}^{(i,j)}$; and from the cells in the same z-column as cell (i, j), $\vec{E}_{same}^{(i,j)}$;

$$\vec{E}_{outside}^{(i,j)} = \vec{E}_{same}^{(i,j)} + \vec{E}_{other}^{(i,j)} + \vec{E}_{ext}. \tag{5.48}$$

$\vec{E}_{other}^{(i,j)}$ must be obtained by summing the $\vec{E}^{(i,j)}(i', j')$ of Eqs. 5.44 and 5.47 over i' and j'. In this section we evaluate $\vec{E}_{same}^{(i,j)}$.

Let $k = 0$ label the cell in the screen plane. $\vec{P}^{(i,j)}$, a vector in the x-y (screen) plane, represents the polarization of all the cells in that z-column so that the field contribution at cell (i, j) of the cell of index k is

$$\vec{E}_k^{(i,j)} = -\frac{\vec{P}^{(i,j)}}{4\pi\epsilon_o k^3} \tag{5.49}$$

(see Eq. 5.38). Thus,

$$\vec{E}_{same}^{(i,j)} = \sum_{k \neq 0} \vec{E}_k^{(i,j)} = -\frac{\vec{P}^{(i,j)}}{2\pi\epsilon_o} \sum_{k=1}^{\infty} \frac{1}{k^3}$$

$$= -\frac{\vec{P}^{(i,j)}}{\epsilon_o} \frac{\zeta(3)}{2\pi}, \tag{5.50}$$

where ζ is the Reimann zeta function,[5] defined as

$$\zeta(n) = \sum_{k=1}^{\infty} \frac{1}{k^n}. \tag{5.51}$$

The specific value needed in Eq. 5.50 is[5] $\zeta(3) = 1.2020569\ldots$.

In the simulation the polarization of all the cells in the (i, j) z-column is altered simultaneously. We must therefore replace Eqs. 5.27–5.31 as follows. (Again the

---

*Their convergence properties are execreble, so much care is needed.

z-column indices are suppressed.) The total electric field in cell (i, j) now consists of four contributions:

$$\vec{E} = \vec{E}_{pol} + \vec{E}_{same} + (\vec{E}_{other} + \vec{E}_{ext}).\qquad(5.52)$$

Using Eqs. 5.26 and 5.50, it can be expressed in terms of the cell polarization:

$$\vec{E} = -\beta\frac{\vec{P}}{\epsilon_o} - \frac{\vec{P}}{\epsilon_o}\frac{\zeta(3)}{2\pi} + (\vec{E}_{other} + \vec{E}_{ext})\qquad(5.53)$$

The cell polarization should be consistent with the total electric field via $\vec{P} = \epsilon_o\chi\vec{E}$, so that

$$\vec{E} = -\chi\left(\beta + \frac{\zeta(3)}{2\pi}\right)\vec{E} + (\vec{E}_{other} + \vec{E}_{ext}),\qquad(5.54)$$

or

$$\vec{E}_{total} = \frac{\vec{E}_{other} + \vec{E}_{ext}}{1 + \chi(\beta + \frac{\zeta(3)}{2\pi})}.\qquad(5.55)$$

Eq. 5.55 determines the electric field in an entire z-column as determined from the mutual effects of the fixed field from other z-columns and external sources and the polarizibility of the cells in the z-column itself. Thus,

$$\frac{\vec{P}}{\epsilon_o} = \chi\vec{E}_{total} = \frac{\chi}{1 + \chi(\beta + \zeta(3)/2\pi)}(\vec{E}_{other} + \vec{E}_{ext}),\qquad(5.56)$$

replaces Eq. 5.31 as the fundamental update equation for a z-column-based iteration. (The numerical value of the factor in the denominator is

$$\beta + \frac{\zeta(3)}{2\pi} = 0.337 + 0.191 = 0.528\qquad(5.57)$$

5.5 Fixed Polarization

In the dielectric materials simulation one is presented with a region of space divided into small cells, 25 in each direction, each of which, if its polarization or susceptibility is set to a non-zero value can represent the presence of dielectric material.* The program also provides for the application of a uniform external field in the horizontal (x-)direction throughout the space displayed.

Within this general framework the simulation is designed with two modes of control: one in which the polarization of each cell is fixed and set by the user; and the other in which the susceptibility, χ_{ij}, is fixed and set. The various physical effects that can be studied in these two modes are outlined in this section and the next, respectively.

---

*If both are set to zero the cell is considered to represent empty space. By turning cells on or off the user controls the shape of the dielectric material to be simulated.

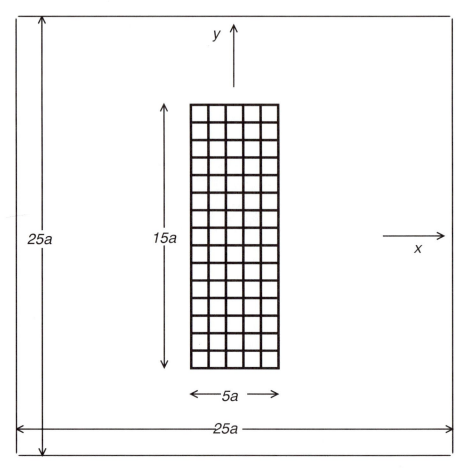

Figure 5.3: A uniformly polarized slab of dimensions $l \times w \times \infty$. (All dimensions are in units of a.)

5.5.1 Uniformly Polarized Slab

Consider first a slab of uniformly polarized material of dimension $l \times w \times \infty$ with $\vec{P}^{(i,j)} = P_o \hat{x}$ as in Fig. 5.3. By the arguments of section 5.3.6 we expect the polarization to result in bound charge densities

$$\sigma_b = \pm P_o \tag{5.58}$$

on the right and left surfaces of the slab. This situation is therefore similar, but not identical to that of a parallel plate capacitor (ppc).* The exercises explore the field due to this uniform slab and the similarities to the ppc.

Another interesting scenario in the fixed polarization mode is a bi-polarized slab, i.e., slabs of two different polarizations in contact. In this case we expect from Eq. 5.32 a bound charge accumulation at the boundary. The presence of this charge can be verified in the DIELECT simulation using Gauss' law.

---

*In a ppc the density of charge of the plates is not constant as we have here on the boundaries of the slab; rather the electric potential is constant on the plates.

5.5.2 Uniformly Polarized Cylinder

Another simple situation with uniformity in the z-direction that can be addressed within the fixed-polarization mode is that of an infinitely long cylinder with constant polarization, $\vec{P} = P_o\hat{x}$. This polarization, using Eq. 5.25 gives rise to a bound surface charge density,

$$\sigma_b = P_o \cos(\phi),\tag{5.59}$$

where the angle ϕ is measured from the x-axis. This charge density gives rise to an electrostatic potential

$$\Phi(\rho,\phi) = \frac{P_o\rho \cos\phi}{2\epsilon_o} \quad \text{for } \rho \le R$$

$$\frac{P_oR^2 \cos\phi}{2\epsilon_o\rho} \quad \text{for } \rho \ge R \tag{5.60}$$

if the cylinder is of radius R. (Verification is left for the exercises.) The resulting electric field can be obtained as $-\vec{\nabla}\Phi$, and it turns out to be uniform within the cylinder:

$$\vec{E} = -\frac{P_o}{2\epsilon_o}\hat{x} \text{ for } \rho \le R.\tag{5.61}$$

This result can be checked with the simulation by setting the polarization of all cells within a radius of R to $\vec{P} = P_o\hat{x}$. Since the simulation is based on Cartesian cells, the cylinder can be only approximately modeled.* (When the **Cylinder** array is selected in the simulation all cells whose centers lie within a distance of $6.1a$ of the center cell are selected.)

Hollow polarized cylinders can also be readily examined similarly with the simulation.

5.6   Fixed Susceptibility

The DIELECT simulation runs in a second mode in which the susceptibility of each cell, $\chi^{(i,j)}$, is fixed and specifiable by the user. This mode simulates the behavior of dielectric materials. This mode is organized similarly to the fixed polarization mode, with the one key difference that the polarization of each cell is iteratively adjusted to convergence using the algorithm specified in section 5.3.3.[†] This section describes some of the situations that can be addressed in this mode.

5.6.1   The Case of One or a Few Cells

The polarization of a single z-column of cells responds to an external field rather like a single atom in the ATOMPOL simulation. The few differences are explored in the exercises.

---

*This is an opportunity to examine the errors induced by the cellular approach.
[†]With the update step altered as given by Eq. 5.56.

5.6.2   Thin Slabs

Thin slabs of polarizable material have more interesting behavior. Let the dimensions of the cell be $1 \times w \times \infty$ with $w \gg 1$ with the thin direction of the slab in the direction of the external field. (Let us term this the *perpendicular* orientation of the slab and the orthogonal one the *parallel* orientation.)

Note first that one does not expect isotropic behavior with respect to slab orientation. In the case of the perpendicular slab, polarization of one cell tends to reduce the field in all other cells, whereas the converse holds in the parallel case. These subjects are explored in the exercises.

5.6.3   The Uniform Cylinder Revisited

We found in section 5.5.2 that the field inside a uniformly polarized cylinder is uniform. This fact, which holds as well for spheres and ellipsoids, can be exploited to determine the polarization of a dielectric cylinder placed into an originally uniform external field.*

Imagine placing an initially unpolarized cylinder, of uniform susceptibility χ, into the region of the external field, $\vec{E}_{ext}$, and visualize its response as a sequence of alterations of its polarization until self-consistency is obtained. The initial response to the uniform external field is a uniform polarization:

$$\frac{\vec{P}_0}{\epsilon_o} = \chi \vec{E}_{ext}. \tag{5.62}$$

This polarization, by Eq. 5.61, results in turn in a new field contribution,

$$\vec{E}_1 = -\frac{\vec{P}_0}{2\epsilon_o}$$

$$\vec{E}_1 = -\frac{\chi}{2} \vec{E}_{ext}, \tag{5.63}$$

which must be added to the original applied field. (Note that the new contribution opposes the original applied field.) This addition will result in a further correction to the polarization

$$\frac{\vec{P}_1}{\epsilon_o} = \chi \vec{E}_1 = -\frac{\chi^2}{2} \vec{E}_{ext}, \tag{5.64}$$

which again alters the field within the cylinder via the addition

$$\vec{E}_2 = -\frac{\vec{P}_1}{2\epsilon_o} = \left(-\frac{\chi}{2}\right)^2 \vec{E}_{ext}, \tag{5.65}$$

again a uniform contribution. This process can be continued to convergence so that the final field is the sum of the $\vec{E}_n'$s deduced above:

---

*The derivative presented here parallels the treatment by Griffiths[1] of the dielectric sphere in a uniform field, and also to some extent models the iterative solution of general dielectric problems employed by the simulation.

$$\vec{E}_{final} = \vec{E}_{ext} + \vec{E}_1 + \vec{E}_2 + \cdots = \left[\sum_{n=0}^{\infty} \left(-\frac{\chi}{2} \right)^n \right] \vec{E}_{ext}. \qquad (5.66)$$

The series is easily summed to

$$\vec{E}_{final} = \left(\frac{1}{1 + \frac{\chi}{2}} \right) \vec{E}_{ext}. \qquad (5.67)$$

(Two facts were critical in the above derivation: the first, that uniform polarization within a cylinder gives rise to a uniform field contribution within the cylinder; and the second, that the polarization is *linear* in the field.) It is left as an exercise to obtain the final polarization of the cylinder and to establish that the solution given is self-consistent.

The result of the above derivation and its breakdown as the cylinder is flattened can be checked using the simulation in the fixed susceptibility mode, as can the response of systems of other geometries. These topics are explored further in the exercise section.

5.7 Exercises

5.7.1 Atomic Polarizability Exercises

5.1 **Atomic Polarization—An Analytic Warmup**
Use the integral form of Gauss' law to verify Eq. 5.6. Assume no distortion of the spherical and uniform charge cloud (of total charge $-Ze$ representing the electrons). (Hint: The force can be determined from the electric field of just the negative charge density at a distance d from the center of its distribution.)

5.2 **Field of an Induced Dipole Moment**
Use the ATOMPOL simulation to investigate the field of an induced dipole as follows. Use the configuration of a single atom located at the center of the screen (the starting configuration) with polarizability set to that of potassium. Set the external field strength to a value of 5. (See section 5.8.2 for a discussion of the units used in the simulation.)

   a. Set the display option to **Combined Field** and describe qualitatively the effect of the polarizable atom on the electric field. Where is the field enhanced and where is it diminished?

   b. Now, using the **Field Probe** tool, make quantitative investigations as follows. Map out the variation of the electric field starting near the atom and proceeding outward along two different directions: parallel to the induced moment and perpendicular to it. Make plots of the field strengths you find versus distance, d, from the dipole and verify graphically that the field decays to the external value like $1/d^3$.

5.3 **Interactions**
Investigate the interactions between polarizable atoms as follows. Using the same external field and polarizability as in exercise 2, place two potassium

atoms on the screen separated by a distance of 10 Å in three different directions: the x-direction, the y-direction, and at a 45 degree angle relative to the x-axis. Compare the resulting polarizations and discuss the differences in each case from the polarization of a single potassium atom in the same external field.

5.4   The Polarization Runaway

Consider two atoms of polarizability α separated by displacement $\vec{d}$ placed in an external field $\vec{E}_o$ such that $\vec{E}_o$ is parallel to $\vec{d}$. In this orientation each atom magnifies the field experienced by the other, so it is possible that the polarizations will increase without limit by amplifying one another if one assumes the simple linear relationship of Eq. 5.1 to govern the induced moments. (Note that this is an *unphysical* limit; the dipole moment cannot expand indefinitely without disrupting the charge clouds of the atoms.) We examine this "polarization runaway" in this exercise.

   a. Show that the magnitude of the polarization of atom 2 can be expressed as

$$p_2 = \alpha\left(E_o + \frac{2p_1}{4\pi\epsilon_o d^3} \right),\qquad (5.68)$$

   and correspondingly for p_1.

   b. Show that these two expressions give

$$p_1 = p_2 = \frac{\alpha E_o}{\left(1 - \frac{2\alpha}{4\pi\epsilon_o d^3}\right)}.\qquad (5.69)$$

   c. What happens to p_1 and p_2 if

$$\frac{\alpha}{4\pi\epsilon_o} \geq \frac{1}{2}d^3 ?\qquad (5.70)$$

   d. Taking the value of $\frac{\alpha}{4\pi\epsilon_o}$ to be that of potassium, determine the distance, d (in Å) at which the polarization should run away. (If one considers closely packed atoms, the result of part c indicates that the divergence occurs when $\frac{\alpha}{\epsilon_o}$ is comparable to the volume taken up by each atom. The nearest neighbor distance in solid potassium is 4.53 Å, so that one is close to this limit. However, due to the alteration of the charge clouds—the outmost electron which results in the large atomic polarizability becomes free of the atom entirely in the formation of the metallic crystal—when the potassium atoms are brought into close proximity, the atomic polarizability is much decreased and thus, excessively large polarization does not occur.)

5.5   Anisotropy

Using ATOMPOL, investigate the response of long atomic chains parallel and perpendicular to the applied field. In which case are the induced moments larger and in which case smaller than those of a single atom in the same field? Explain the result.

5.6 Fields at Other Angles

The menu of available atomic configurations in ATOMPOL includes a **Diagonal Line** of atoms. Consider the behavior of such an array of atoms for the potassium polarizability and a variety of field strengths. Discuss the angle of the induced dipole moments. Why is the induced moment not parallel to the field in this case? Does it vary with field strength? Next repeat this experiment with all atoms assigned a smaller polarizability. How does this affect the angle of the induced moments?

5.7 Dielectric Forces on Atoms

Polarizable atoms in close proximity not only influence one another's polarization but also exert forces on one another.

a. Is the force between atoms due to their polarization attractive or repulsive? Does the orientation of the vector separating them relative to the external field make a difference? Use the ATOMPOL program to verify your answer by placing two atoms of the polarizability of potassium at various orientations relative to the external field of strength 5.

(Hint: A uniform field exerts no net force on a dipole, but there is a force when the gradient of the field is nonzero. See Eq. 5.5 in Griffiths,[1] for example. How can you use the field probe to estimate the magnitude of the electric field gradient in the direction of the dipole moment of an atom? Explain the procedure you will use. Note that the field probe does not include the field due to any atom within a distance of 2.5 Å from it. Thus if one clicks inside one of the atoms, the field obtained does not include the contribution of that atom.)

b. Dielectric solids can be modeled as collections of many polarizable entitles in close proximity. Do the interactions considered in part a lead to a compressive or expansive strain on dielectric materials?

5.8 Atomic to Bulk

Dielectric materials can be modeled as collections of polarizable entities, much like the atoms of the ATOMPOL simulation. The transition from atomic to bulk behavior can be addressed by considering clusters of atoms of various sizes. Using $\alpha = 10$, investigate the polarization response of 1-atom, 4-atom, 9-atom, 16-atom, and 25-atom square clusters. (The 1-atom and 25-atom cases are available as preset configurations; you can input the others using the **Keyboard Atom Plcmnt** tool. How does the typical polarization depend on cluster size? Speculate on the limiting behavior for $n \times n$ clusters as n gets large.

5.9 Crystal Structure Dependence

The available configurations include three arrays that mimic two-dimensional solids of varying crystal structure. They are **Square Lattice, Rectangular Lattice,** and **Close-Packed Lattice**. Using $\alpha = 10$, investigate the response of these three structures to an external field of strength 5.0 units. In which case is the average polarization largest? In which is it the smallest? Can you explain the results? (Hint: You will want to use the **Probe Atom** tool to investigate the polarization of several non-edge atoms in each case.)

5.10 The Polarization Runaway II

The polarization runaway can be observed for clusters of atoms. Take the polarizability to be that of potassium and choose the **Close-Packed Lattice** array. Run the polarization solution and describe what occurs.

5.7.2 Fixed Polarization Exercises

The following exercises explore the behavior of dielectric materials using the DIELECT simulation in the fixed polarization mode.

5.11 The Dipole Field in Cartesian Coordinates

Show that Eq. 5.38 gives the electric field of a dipole moment with $\vec{p} = (p_x, p_y, 0)$, i.e., no z-component. Hint: Start from the expression for the potential of a dipole (Eq. 5.2) rather than that for the field in spherical coordinates (Eqs. 5.3, 5.4).

5.12 Field of a Polarized z-Column

With the external field set to 0 and just a single active polarized cell with $\vec{P}/\epsilon_o = 10\hat{x}$, determine the resulting electric field as a function of distance d from the cell separately in the x- and y-directions. (In so doing you are investigating the dependence of $S_1(\Delta_x, 0)$ on Δ_x and $S_1(0, \Delta_y)$ on Δ_y.)

5.13 Field of a Uniformly Polarized Slab

Using the DIELECT simulation, set up a slab of width $w = 5$ and length $l = 15$ (as in Fig. 5.3) with fixed polarization $\vec{P}/\epsilon_o = 10\hat{x}$. Set the external field to 0.

a. Imagine a Gaussian surface in the shape of a cube with edge length a placed such that it is located at the center of the right-hand surface of the slab and is bisected by it. Use the field probe to determine the flux of electric field on the right and left surfaces of the cube. Using Gauss' law, determine the net charge that must be enclosed by this Gaussian surface. (Hint: See section 5.8.3 for a brief discussion on the use of Gauss' law in the cell-based simulation.) From this determine the bound surface charge density σ_b on the right-hand surface of the slab. Is this value consistent with the polarization of the slab?

b. Again using the field probe, determine the dropoff of the field due to the slab as a function of distance from the edge of the slab in the y-direction. Make a plot of E_x vs. y for y ranging from 0 to 20 in units of a. What power law governs the drop off of the field strength?

c. Examine the electric displacement $\vec{D}$ associated with your polarized slab. In this situation there is no free charge density anywhere, so that by Eq. 5.36 one might expect $\vec{D} = 0$ everywhere in space. Since that is evidently not the case, you need to explain what is happening. Hint: See section 5.3.2 of Griffiths.[1]

5.14 Comparison of Fixed Polarization Slab to Capacitor

The situation of the uniformly polarized slab resembles that of a parallel-plate capacitor (ppc) of similar dimensions. The difference is that in the case of the slab the charge density on the surfaces is uniform, while in the case of the ppc, the electric potential along the surface is constant. Explore this difference by setting up a slab like that of the preceding exercise, but with $w = 1$.

a. Using the field probe, determine the approximate electric potential difference between the center of the right-hand surface of the slab and its upper right end. (Use the $\vec{E}$-field values in the empty cells immediately to the right of the right surface of the slab for this determination.)

b. This potential difference will vanish if the cell polarizations can be adjusted so that the field in all the cells immediately to the right of the slab is perpendicular to the surface of the slab. In which direction (i.e., increase or decrease) do the end cell polarizations need to be adjusted to reduce this potential difference? Test your hypothesis by either doubling or halving the end cell polarizations and repeating part a. What does this study tell you about the arrangement of charge on the plates of a parallel plate capacitor?

5.15 Field of Uniformly Polarized Cylinder

Consider a uniformly polarized, infinitely long cylinder.

a. Verify that Φ given in Eq. 5.60 satisfies Laplace's equation for $\rho \neq R$.

b. Staying in cylindrical coordinates, determine the electric field consistent with this potential inside and outside the cylinder. Show that the field you obtain inside the cylinder is consistent with Eq. 5.61.

c. Using Gauss' law show that the field discontinuity at $\rho = R$ is consistent with the bound charge density specified in Eq. 5.59. (Hint: See section 5.8.3 for a brief discussion on the use of Gauss' Law in the cell-based simulation.)

5.16 The Uniformly Polarized Cylinder

Set up in the simulation an infinitely long cylinder (axis in the z-direction) with fixed polarization $\vec{P}^{(i,j)}/\epsilon_o = 10\hat{x}$. (The choice **Cylinder** under menu heading **Configs.** sets all the cells with $(i^2 + j^2)^{1/2} \leq 6$ to the specified polarization.)

a. Using the field probe determine the electric field in all the cells within the cylinder. Is the field constant and in the correct direction? Discuss any deviations from expectations. Extra credit: Investigate the effect of the discrete grid in this case by studying the convergence of the field strength to a constant value as one moves from the edge of the cylinder

inward. Show that the convergence is approximately Gaussian. Double
extra credit: Create a larger cylinder, say, of radius 8 instead of 6, by
turning on the appropriate cells around the periphery of the radius 6 cylinder. Investigate the covergence in this case as well. (Approximately
the same dependence occurs in this case as for $r = 6$.)

b. Again, using the field probe, determine the dropoff in magnitude of the
field as a function of distance away from the cylinder in various directions. Do you find the $1/\rho^2$ dependence suggested by the results of the
preceding exercise?

c. Examine the electric displacement $\vec{D}$ in this situation. There is no free
charge or external fields, yet $\vec{D}$ does not vanish. Why not?

5.7.3   Fixed Susceptibility Exercises

The following exercises explore the behavior of dielectric materials using the
DIELECT simulation in the fixed susceptibility mode.

5.17   A Single Cell

With the simulation in the fixed susceptibility mode and the external field
set to 5, turn on a single cell (which represents an entire z-column of cells,
of course) and set its susceptibility to be $\chi = 2$. Start the self-consistent determination of the polarization of the cell using hot key F2.

a. One might have thought on the basis of Eq. 5.15 that the polarization
of the cell would turn out to be $\vec{P}/\epsilon_o = 10.0$. What value does the simulation obtain? Explain why the answer is not that of the naive specula-
tion. (Hint: See section 5.4.2.)

b. If this were a polarizable atom, the alteration of the electric field in the
vicinity of the polarized cell would "heal" back to the external value
like $1/r^3$, where r is the distance from the cell. Check the field values
in five cells going away from the polarized cell in an upward direction,
i.e., orthogonal to the external field. Record the values and check the
approximate power dependence on r by making a log-log plot. Why do
you find a power smaller than 3?

5.18   Adjacent Pairs of Cells

Set up the simulation such that two adjacent cells, (i, j) and $(i + 1, j)$ (so
that the cells are neighbors in the x-direction), have been turned on with
susceptibility χ set to 2. Run the polarization iteration to obtain the self-
consistent solution in this case.

a. Compare the polarization of the two cells to that found for a single cell
in the preceding exercise. Explain why the polarizations are larger.

b. Now change the cells such that (only) two cells that are neighbors in the
y-direction are turned on in the same fashion as above. Compare the resulting solution to that obtained in part a and explain any differences.

5.19 The Dielectric Cylinder in a Uniform Field

In section 5.6.3 we found the field within the polarized cylinder by a method of successive approximations. Let us investigate the uniform cylinder by means of the simulation. Select the **Cylinder** configuration and run the calculation to determine its polarization in response to an external field of value 5.

a. Determine the final polarization of the cylinder and show that it is consistent with Eq. 5.15.

b. Verify that the polarization within the cylinder is uniform by examining (using the **Field Probe**) its variation as one moves away from the center of the cylinder in both the *x*- and *y*-directions.

c. What is the effect of the cylinder on the field in the region immediately adjacent to it? Where is the external field strengthened and where is it weakened?

d. Now "flatten" the cylinder by removing a row of cells (symmetrically) from its top and bottom. Run the polarization calculation again and discuss the results. If the field is no longer approximately uniform within the cylinder identify where it is stronger and where weaker. Explain this pattern.

5.20 Depolarization

A thin dielectric sheet oriented perpendicular to the applied field polarizes to a lesser extent than one oriented parallel to the field. The DIELECT simulation offers **Thin Parallel** and **Thin Perpendicular** configuration options to conveniently study this effect.

a. Determine the polarization of the middle cell of the sheet in both (thin slab parallel to or perpendicular to the external field) cases. Set the susceptibility of all cells to $\chi = 4.0$ and the external field to a value of 5.

b. In which case is the effect on the surrounding region, i.e., the deviation of the field strength from the externally applied value, greater?

c. Explain your result in part a.

5.21 Convergence of the Iteration

To observe the convergence in the iterative calculation, it suffices to consider situations with just two active cells. Using the step-by-step version of the fixed simulation mode (see section 5.8.3), observe the convergence of the polarization of pairs of cells:

a. oriented end to end (adjacent in the *x*-direction); and

b. oriented side-by-side.

In each case observe the step-by-step convergence of cells with $\chi = 4$ in an external field of strength 5. In which case is the convergence oscillatory? In which is it monotonic? Explain these results.

c. In both of the above cases the resulting polarization is in the *x*-direction. If the two cells are diagonal neighbors, the polarization is not along the axis. Run this case to determine the direction of the polarization and explain why the polarization cannot lie along the axis.

5.22   Ferroelectrics

Consider Eq. 5.23. What happens to the susceptibility of a material if the polarizability of the entities of which it is composed is such that $n\alpha/(3\epsilon_o) \simeq 1$? (Hint, see exercise 4.) Speculate on the behavior of a material in which that condition holds. (Such materials do exist; the ceramic $BaTiO_3$ is an example.)

5.23   The Breakdown of Stokes' Theorem

Another measure (see ex. 16) of the errors induced by the discrete grid is afforded by Stokes' theorem, which implies that since in any static situation $\vec{\nabla} \times \vec{E} = 0$, the line integral of the electric field about any closed path should vanish,

$$\oint_C \vec{E} \cdot \vec{dl} = 0, \tag{5.71}$$

where C is any closed contour. All the electric fields included in the simulation, the uniform external field as well as any induced-dipole fields are curl free.

a. Show that the dipole field (Eqs. 5.3, 5.4) is curl-free for all $\vec{r}$, except at the origin.

Therefore Eq. 5.71 should hold in the simulation. However, due to the fact that the field within a polarized cell is taken to be uniform, whereas outside that cell, the field produced by its polarization is described by the correct functional forms of Eqs. 5.3 and 5.4, there will be considerable inaccuracies, particularly near the boundaries of dielectric materials. Investigate this breakdown as follows.

b. Using the fixed susceptibility mode, choose the **Parallel Slab** configuration with the external field set to a value of 5 and susceptibilities of all cells in the slab set to a value of 4. Run the iteration to obtain the self-consistent solution and then, using the **Field Probe**, record the *x*- and *y*-components of $\vec{E}$ for the following eight cells: $(13, 14)$, $(13, 15)$, $(13, 16)$, $(13, 17)$, $(14, 14)$, $(14, 15)$, $(14, 16)$, and $(14, 17)$.

These cells can be used to check Eq. 5.71 around three different small square contours: one inside the slab, $C_1 = (13, 14) \rightarrow (14, 14) \rightarrow (14, 15) \rightarrow (13, 15) \rightarrow (13, 14)$; one outside the slab, $C_2 = (13, 16) \rightarrow (14, 16) \rightarrow (14, 17) \rightarrow (13, 17) \rightarrow (13, 16)$; and one that lies on the boundary of the slab, $C_3 = (13, 15) \rightarrow (14, 15) \rightarrow (14, 16) \rightarrow (13, 16) \rightarrow (13, 15)$.

c. Do the line integral defined in Eq. 5.71 for each of the three contours. Assume the field components vary linearly between nodes of the grid.

For which contour(s) is Stokes' theorem approximately satisfied? For which is there big trouble?

(N.B. Stokes' theorem can be used to argue that the parallel component of $\vec{E}$ is continuous across dielectric boundaries. Given its breakdown in our cell-based approach, the simulation does not accurately reproduce this expectation.)

5.7.4 Possible Modifications to the Program

There are a number of extensions that can be made to this model to improve and broaden the physics content. The following subsections offer a partial list of possible extensions in the two program modes. Those marked with an asterisk should be undertaken only by students with experience in Pascal and object-oriented programming.

Atomic Polarization

5.24 **Increased Number of Atoms**
The ATOMPOL program is written with a limitation of the number of atoms to 25, mostly for convenience in labeling the atoms with single letters of the alphabet in screen displays. It is straightforward to increase the limit on the number of dipoles, simply by increasing the constant **Max-Atoms** near the top of the program and by increasing the dimension of **TYPES DipoleVal** and **DipoleBool** so that the array dimensions are correspondingly increased. Note that the program does not allow atom overlap within the region depicted in the main window, so that the number of atoms cannot be increased without limit.

5.25 **The Polarizability Tensor***
The ATOMPOL program is written to treat only spherically symmetric polarizable entities, i.e., those whose polarization response is always in the direction of the applied field. Real, nonsymmetrical molecules respond anisotropically, as discussed, for example, by Griffiths[1] in his section 4.1.2, where the relevant polarizability information is given for CO_2. To handle this case one would need to add arrays specifying the new elements of the polarizability tensor for each atom and the orientation of the molecular axis of each. (Both can conveniently use the **TYPE DipoleVal** defined in the program.). The appropriate generalization of Eq. 5.1 is

$$\vec{p} = \alpha_{\perp}\vec{E}_{\perp} + \alpha_{\parallel}\vec{E}_{\parallel}, \tag{5.72}$$

where $\alpha_{\perp}$ and $\alpha_{\parallel}$ are, respectively, the polarizabilities of the molecule perpendicular to and parallel to its axis.

5.26 **Three Dimensions***
All atoms are restricted the the *x-y* plane in the current simulation. It would be straightforward to allow atoms to be at locations with non-zero *z*-component. Obviously changes will be required throughout the program and should be attempted only by students willing to invest the time required

to understand the details of the user interface. A second complication is the matter of display. One could choose to display atoms and dipole moments (but not fields!) projected onto the *x-y* plane. For more details on the 3-D case, see the next section.

Dielectric Media

5.27   High Susceptibility Materials

The maximum susceptibility allowed by the program is 10. In ex. 22 above, the significance of very high susceptibilities is noted. It is straightforward to alter the maximum susceptibility allowed in the program to rather large values. This is done by altering the variable **slider2max** in the procedure **DefineSliders.** N.B.: Make the change only for **Dielectric-PolMode<>1**, the fixed susceptibility mode.

If you make this modification, it is interesting to make a plot of the variation of the polarization of a slab of high susceptibility as a function of the applied external field.

5.28   Line Charges

The simulation applies an external field in the *x*-direction only. For dielectrics of cylindrical shape, it is also interesting to consider the application of field that radiates outward from the center of the material as if there were a line charge present there. It is straightforward to alter the program to accomplish this change. The effect of the external field occurs only in a single statement in the procedure **DielectricEfield**. Do not forget to write in the appropriate dependence of field strength on distance from the center. (The central cell is indexed as **wi = 13** and **wj = 13** in the program.)

5.29   Three Dimensions*

It would not be prohibitively difficult to modify the program to treat 3-D situations. The changes that need to be made include—

- The alteration of the field and susceptibility arrays to be 3-D. This is most conveniently done by changing the **TYPE**s **DielectricArray** and **DielecBoolArray**. New arrays for the *z*-components of the electric field and polarization will be needed and must be initialized in **InitializeGlobals** then altered by procedure **EraseArrays**, which clears all arrays. It would be natural to name them **ElcZ** and **PolZ**, respectively. Wherever the field or susceptibility arrays are specifically indexed, they would of course acquire the third index.

  The size of the region simulated is (25 × 25) in the present program and the central cell is indexed (13, 13). It would be convenient to use length 25 also for the *z*-direction and to regard the central cell to be that with indices (13, 13, 13).

- An additional loop for the *z*-direction will need to be nested within loops for the *x*- and *y*-directions (and in some cases the inclusion of *z*-index ranges in the argument lists) in the following procedures:

FillDielectricArray, FillDielectricBoolArray, FillSelected-DielectricArray, FindSumLimits, EfieldSum, SolvePolarization, and **Initialize**, and in the hot key handling part of the the main program.

- One would have to worry a bit about display considerations. It would be easiest to stay with a 2-D display on the screen, with perhaps the plane of display under the users control via an input screen (see CUPS utilities manual). Field vectors could be drawn in 2-D projection onto the screen plane (thus allowing one to avoid changes in the procedures **DrawArrow** and **DrawCentArrow**). The third component of all vectors could be provided by writing its value in the closeup window, requiring changes in procedure **DrawCellCloseup.**

- The procedure that generates the default configurations, **DielectricArrayHandler**, will need to handle the z-direction as well. This will require corresponding changes in **SelectArray**.

- One of the few changes in the physics would include expressing the dipole field in 3-D and making the appropriate changes in the procedures (**EfieldSum** and **DielectricEfield**) that determine the field in a given cell or throughout the region due to all other cells and the external field. Note that the Reimann sum factors, **PolSumX** and **PolSumY**, are no longer needed. The arguments of the procedure would include the z-index of the cell, **wk** presumably, and it would have to also return the z-field component, **NewEZ**.

- The self-field correction based on Eq. 5.56 in procedure **SolvePolarization** should be altered to be consistent with Eq. 5.31.

- If anyone accomplishes this extension the author would very much appreciate receiving a copy of the program (**LROELOFS@ HAVERFORD.EDU**).

5.8 Running the Programs

To run either of the two programs associated with this chapter simply type the name, ATOMPOL or DIELECT, respectively. The programs come to life displaying a credit screen; clicking the mouse or pressing any key initiates the program.

5.8.1 General Introduction

The CUPS simulations are all written with a rather self-explanatory user interface and you will probably be able to explore the programs without much assistance. The student is referred to Chapter 1 of this volume for general instructions on using the CUPS simulations and to the help screens available in the program for more particular assistance.*

---

*The help utilities are only available if the relevant .HLP files are located in the directory in which the simulation program resides. If the file is missing, check with your instructor.

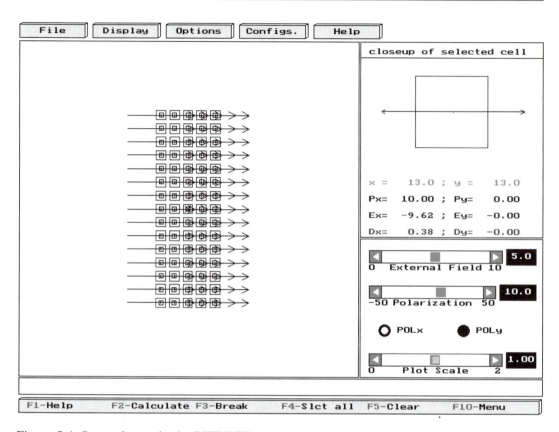

Figure 5.4: Screen layout in the DIELECT program. In the main window, the inactive cells are not visible. The cell in the main viewport labeled with an "x" is displayed in the closeup window. The arrows in the cells in the main viewport represent the polarizations of individual cells. Alternatively $\vec{E}$ or $\vec{D}$ arrows may be displayed throughout the entire region. In the ATOMPOL program electric field lines are drawn instead of cell-based arrows.

Figure 5.4 shows the appearance of the screen of the DIELECT program. The screen for the ATOMPOL program is similar except that the main viewport displays the polarized atoms and the closeup window is the output window for field probe like that of the IMAG&MUL program of Chapter 4 or a cartoon display of the polarization of a selected atom.

More-or-less self-explanatory hot keys are available in both programs and displayed at the bottom of the screen. These may be activated either by clicking on them with the mouse or by hitting the associated function key on the keyboard.

The programs have different main menu bars. Both include a file menu for controlling the overall operation; the menu items under the **FILE** menu are as described in chapter 4. The program-specific menu items and the operation of the two programs are described in the following subsections.

5.8.2   ATOMPOL

The ATOMPOL simulation treats the effect of external fields on polarizable entities (to be called *atoms* in the following) and their mutual interactions. The interaction of the atoms via their dipole fields is treated self-consistently in the program
via an interative approach.

Display and Control Issues

The program allows placement of up to 25 atoms in a two-dimensional region of
space that is 25 Å × 25 Å in size. This region is shown in the main window of the
simulation and the atoms are displayed as circles in this area.

The strength of the applied external field (always in the *x*-direction) and the
atomic polarizabilities are controllable via sliders. (The polarizability slider
changes that parameter only for the *selected* atoms. Atoms are selected, or de-
selected, by means of mouse clicks. Selected atoms are displayed in magenta and
de-selected atoms in the color cyan.)

Atom locations can be specified in a number of ways: via a choice among eight
different default configurations on a menu, via an input screen, or by double-
clicking with the mouse at the desired point (to add an atom) or on a chosen atom
(to remove it). The hot keys (displayed at the bottom of the screen) add to the ease
and flexibility of manipulation of the atoms. Hot key (function key) F1 produces
a help screen; F3 causes all the atoms on the screen to be selected; F4 clears all
selections (but does not remove the atoms); and the **Delete** key deletes all selected atoms, i.e., removes them from the screen. To delete *all* the atoms, whether
selected or not, choose **Clear All** under the **Configs.** heading at the top of
the screen (which should not to be confused with the **Config.** option under the
File menu).

The self-consistent calculation described in section 5.2.2 does not occur automatically whenever the atoms are added to or removed from the region. It occurs,
rather, only when the hot key F2 is pressed. The iteration is run for a maximum
of 25 loops through all the atoms present. This is enough to obtain convergence
for most arrangements of atoms. If convergence does occur the user is notified via
a message at the bottom of the screen. If convergence is not obtained, the calculation may be continued for another 25 iterations by pressing F2 again. During the
iteration the calculation may be interrupted by pressing the F2 key. In that event
the dipole moments found during the last sweep through the atoms are used in
subsequent displays, etc.

Several tools for determining the field at any point or the characteristics of
any of the visible atoms are provided and also explained below.

Units

The screen represents an area 25 Å × 25 Å, and the program uses Å as the distance unit throughout. All atoms are drawn as circles of diameter 5 Å.

All polarizabilities used in the program are actually $\alpha/4\pi\epsilon_o$ and are given in
units of Å^3. (See Eq. 5.10.) Given Eq. 5.1, it is convenient in the program to work

in terms of and display $\vec{p}/4\pi\epsilon_o$ rather than $\vec{p}$ itself. The quantity $\vec{p}/4\pi\epsilon_o$ has natural dimensions of $[fieldunits] \times \overset{\circ}{A}^3$, see Eqs. 5.3 and 5.4 since distances are in angstroms. It therefore remains only to specify the field units. If the electric fields are taken to be in $\mu V/\overset{\circ}{A} = 10^4 V/m$, one has a reasonable situation both macroscopically and microscopically. The allowed external fields range from 0 to 10 in these units. If we adopt $\mu V/\overset{\circ}{A}$ as the field unit (the simulation uses a dimensionless field), then $\vec{p}/4\pi\epsilon_o$ is in units of $\mu V \overset{\circ}{A}^2$. The characteristic size of this quantity in the simulation ranges up to $10\mu V/\overset{\circ}{A} \times 34 \ \overset{\circ}{A}^3 = 340 \ \mu V \overset{\circ}{A}^2$. In drawing dipole arrows on the screen a scale factor is applied so that maximum expected dipole moment of an isolated atom will plot as an arrow of somewhat shorter length than the atom diameters.

Menu Items

The menu options available in the ATOMPOL program are as follows.

- **DISPLAY**: Controls display options for the main window.

    - **External Field**: External applied electric field is displayed. (All electric fields are displayed in yellow with arrows to indicate direction.) The strength of the field is controlled by the slider; it is always directed in the x-direction (horizontal, on the screen).

    - **Single Dipole Field**: Display the electric field due to a single polarized atom, the last one selected.

    - **Combined Field**: Display the total electric field, that due to the external source, and all polarized atoms on the screen.

    - **Label Atoms w/ Dipole Arrows**: All displayed atoms are labeled with an arrow (in white) representing the magnitude and direction of their dipole moments.

    - **Label Atoms w/ Letters**: All displayed atoms are labeled with lowercase alphabetic characters for unique identification.

- **TOOLS**: Controls the operation of program utilities and display options for the other windows on the screen.

    - **Selected Atom (Nuclear Scale)**: The upper right window shows a cartoon sketch of a typical atom whose polarizability is that of the current slider setting immersed in an external field, of magnitude equal to that current slider setting. The shift of the nucleus relative to the electron charge cloud is depicted on the scale of typical nuclear dimensions.

    - **Selected Atom (Atomic Scale)**: The upper right window shows a cartoon sketch of a typical atom (see above) drawn at atomic scale.

    - **Atom Input Screen**: Opens a viewport for specification via keyboard of the number of atoms and their locations on the screen. (Atoms can also be added to or deleted from the display using the mouse and hot keys as noted above in this section.)

– **Field Probe**: The *field probe* is a tool which determines and displays in the upper right window the magnitude and direction of the $\vec{E}$-field at any particular point. Choosing this option activates the field probe. Clicking with the mouse anywhere in the main window results in display of the location of the mouse and the field evaluated at that point. (Field is given in both Cartesian representation and in the form of magnitude and direction.) Also an arrow representing the magnitude and direction of the field is drawn (in blue) at the point of the click. To exit from the field probe hit any key or click outside the main window.

– **Probe Selected Atom**: Displays polarization and other information concerning the last selected atom in the upper right window.

– **Refresh Screen**: Selection of this item causes the screen to be cleared and redrawn. (This eliminates the blue arrows drawn by the field probe.)

● **Configs.**: Menu of available pre-programmed arrangements of atoms; names are self-explanatory.

– **Single Dipole**: This is the default configuration.

– **Horizontal Line**

– **Vertical Line**

– **Diagonal Line**

– **Square Lattice**

– **Rectangular Lattice**

– **Close-packed Lattice**

– **Alloy**

● **HELP**: A pull-down menu of help screens relating to other menu items.

5.8.3   DIELECT

Display Issues

The DIELECT program simulates the behavior of dielectric materials in applied fields using a cell-based approach. The main display area can be considered to be divided into a (25×25) array of cells, each of which may represent either empty space or a small bit of dielectric material. Dielectric materials of different shapes are simulated by activating groups of cells. A cell is said to be "active" if it contains dielectric material. Active cells are outlined in the main viewport using a color code. Inactive cells, which represent empty space, are not visibly displayed in the main viewport.

The DIELECT program runs in two different modes: one in which the polarizations of the cells are fixed and controlled by the user; the other in which the susceptibilities of cells are fixed. In the fixed polarization mode all active cells are outlined in blue. In the fixed susceptibility mode a color scheme with increasing intensity represents increasing susceptibility. (The **Cell Display** help screen in the program gives the details.)

Studying dielectric materials of different shapes requires the facility of altering the polarization or the susceptibility of individual or groups of cells. This can be accomplished using the slider that controls the parameters of "selected" cells. Selected cells are indicated via a small white inscribed square. One can select or deselect individual cells via a mouse double-click. The hot key F4 selects all active cells in the viewport; F5 deselects all active cells. The polarization or susceptibility values of selected cells are not altered until the sliders are next adjusted. The menu item **CLEAR** under **ARRAYS** on the main menu bar clears the screen entirely so that no cells are active or selected.

The strength of the applied external field can be controlled using another slider. This field always acts in the x-direction (horizontal on the screen).

Units

The spatial dimension of an individual cell is taken throughout this writeup to be a. The program DIELECT is set up in a dimensionless fashion so that a is taken to be unity.

The fields $\vec{E}$, $\vec{P}/\epsilon_o$, and $\vec{D}/\epsilon_o$ can be seen to have identical units. Again the program is set up dimensionlessly so that the three fields are simply given and referred to numerically. One may apply any unit desired; the choice volts/a is probably best, bearing in mind that it will simplify the determination of potential differences.

Gauss' Law

Some of the exercises involving the DIELECT program suggest the application of Gauss' law to determine bound charge densities on various surfaces. It is intended that the *integral form* of the law, which relates the integrated electric flux on a closed surface to the net charge enclosed by the surface, be used in these contexts. This is done as follows. The screen display in the DIELECT program includes a closeup window which specifies *inter alia* the electric field of any cell (chosen via a mouse single click). One readily imagines small cubical Gaussian surfaces whose faces coincide with the centers of the cells. Assuming constant field strengths throughout the cells allows for convenient approximation of the flux "integrals."

Menu Items

The menu headings available in the DIELECT simulation are as follows.

- **DISPLAY**: Selects between fields for display. The selected field is displayed in the form of an arrow proportional to the strength of the field, drawn centered in the cell in which it has been evaluated. Note that for consistency of (SI) units, the program actually calculates and displays $\vec{P}$ and $\vec{D}$ divided by ϵ_o.

  – **Polarization**: Displays the polarization field $\vec{P}/\epsilon_o$.

  – **Electric field**: Displays $\vec{E}$.

  – **Displacement**: Displays $\vec{D}/\epsilon_o$.

- **OPTIONS**: Controls the program mode, fixed polarization, or fixed susceptibility.

  - **Refresh Screen**: Redraws main viewport and closeup window.

  - **Fixed Polarization**: Puts program in fixed polarization mode. In this mode cell polarizations are under control of user.

  - **Fixed Susceptibility (Iterate to Convergence)**: Puts program in the fixed susceptibility mode—cell susceptibilities are controlled by user. All iterative solutions (initiated with the F2 hot key) of the resulting polarizations are carried through to convergence or a maximum of 20 steps.

  - **Fixed Susceptibility (Iterate Step by Step)**: Also puts program in the fixed susceptibility mode, but in this case the F2 hot key initiates a single iteration in the polarization solution. This choice is useful for observing the convergence in detail.

- **Configs.**: This selection provides a menu of pre-programmed dielectric material shapes. Other shapes you wish to investigate must be entered via mouse manipulations as described above in this section. All dimensions are given in units of a, the basic cell size. All shapes are assumed to be infinitely long in the z-direction.

  - **Perpendicular Slab**: (15×5) slab with long direction perpendicular to the direction of the applied external field, $E_{ext}\hat{x}$.

  - **Parallel Slab**: (15×5) slab with long direction parallel to the direction of the applied external field.

  - **Cylinder**: Cylinder of radius $R = 6.1$. A cylinder cannot be represented smoothly as a collection of Cartesian cells. It is approximated by activating all cells whose centers lie within a distance R of the center of the screen.

  - **Thin perpendicular Slab**: (15×1) slab with long direction perpendicular to the direction of the applied external field.

  - **Thin Parallel Slab**: (15×1) slab with long direction parallel to the direction of the applied external field.

  - **Double Perpendicular Slabs**: Adjacent (15×3) slabs with long direction perpendicular to the direction of the applied external field arranged a distance 3 units apart.

  - **Clear**: Deactivates all cells on the screen. This selection is useful for emptying the screen if one wants to consider the behavior of one or just a few cells, which can then be conveniently activated using mouse clicks.

- **HELP**: This selection provides help screens relating to other menu items.

Acknowledgments

Nathanial Johnson, Haverford class of 1993, did much of the original programming for this simulation, contributed many insights, and taught me Pascal. My colleague Jon Marr suggested the exercise on Stokes' theorem.

References

1. Griffiths, D. J. *Introduction to Electrodynamics*. 2nd ed. Englewood Cliffs, NJ: Prentice Hall, 1989.

2. Reitz, J. R., Milford, F. J., and Christy, R. W. *Foundations of Electromagnetic Theory*. 3rd ed. Reading, MA: Addison-Wesley, 1979.

3. Lorrain, P., Corson, D. R., Lorrain, F. *Electromagnetic Fields and Waves*, 3rd ed. New York: Freeman, 1988.

4. Styer, L. *Biochemistry*. 3rd ed. New York: Freeman, 1988.

5. Abramowitz, M. and Stegun, I. A. *Handbook of Mathematical Functions*. New York: Dover, 1965.

6

Magnetostatics

Ronald Stoner

> I experienced a miracle, as a child of age 4 or 5, when my father showed me a compass.

> —Albert Einstein

6.1   Introduction

Magnetostatics is the study of magnetic fields and magnetic forces produced by steady electric currents. The study of magnetostatic phenomena dates back to experiments with lodestones and magnetic compasses, but we now understand that magnetism and electricity are closely related phenomena. The magnetic force exerted by one moving charge on another moving charge in one frame of reference would appear as a purely electrostatic force between the same charges in a different reference frame where one or both of the charges are not moving. In that sense, we can think of a magnetic field as being a consequence of an electric field. However, it is equally valid to think of the electric field observed in one reference frame as being a consequence of the magnetic field observed in a different frame. It is better yet to think of magnetostatic and electrostatic fields as each being a special case of the more general phenomenon of electromagnetic fields.

The above explanation of magnetism is easy to accept when the magnetic fields are obviously associated with moving charges and electric currents, such as in electromagnets. The magnetism of lodestones and other magnetic materials may seem more mysterious, but that magnetism, too, is a consequence of the motions of charges. The behavior of magnetic materials is the result of steady electric currents on atomic and subatomic scales that are associated with the perpetual quantum mechanical orbital motions and intrinsic spins of electrons.

6.2   The Basic Laws of Magnetostatics

Maxwell's equations are the basic laws of electromagnetism, and they fully satisfy the covariance requirements of special relativity, namely that they have the same mathematical form in all inertial frames of reference. Maxwell's equations apply to magnetostatics when both the electric current density and the electric charge density are independent of time. When charge and current are static, the electric and magnetic fields in space are also static, and the magnetic fields can be treated separately from the electric fields. In this static case, Maxwell's equations for the electric field $\vec{E}$ and the magnetic induction field $\vec{B}$ are

$$\vec{\nabla} \cdot \vec{E} = \frac{\rho}{\epsilon_o}, \tag{6.1}$$

$$\vec{\nabla} \times \vec{E} = 0, \tag{6.2}$$

$$\vec{\nabla} \cdot \vec{B} = 0, \tag{6.3}$$

$$\vec{\nabla} \times \vec{B} = \mu_o \vec{J}. \tag{6.4}$$

Here $\vec{B}$ is measured in tesla, $\vec{J}$ is in amperes per square meter, ϵ_o is called the "permittivity of free space," and μ_o is called the "permeability" of free space. The first two of Maxwell's equations are the differential forms of the basic laws of electrostatics, and the second two are the differential forms of the basic laws of magnetostatics.

The equation $\vec{\nabla} \cdot \vec{B} = 0$ has a simple physical interpretation: Since the divergence operator (i.e., $\vec{\nabla} \cdot$) applied to a vector field gives the density of its sources and sinks (density of origins and terminations of the field lines), this equation requires that magnetic field lines always form closed loops in space. This contrasts with the case of electrostatic fields, which always originate and terminate on electric charges. Therefore, there is no analogue in magnetostatics of the electric charges of electrostatics. That is, there are no "free magnetic charges" or "magnetic monopoles" in Nature.

6.2.1   The Biot-Savart Law

The relationship between magnetic fields and the electric currents that produce them was understood in the early 19th century through the work of French physicists J. B. Biot, Felix Savart, and A. M. Ampere. The result of their work was the formulation of a law for computing the contribution $d\vec{B}$ to the magnetic field at a point $\vec{r}$ in space of a small "current element" $I\,d\vec{l}$ located at some other point $\vec{r}'$. The form of the Biot-Savart law in rationalized mks units is

$$d\vec{B} = \frac{\mu_o}{4\pi} \frac{I\,d\vec{l} \times \vec{R}}{R^3}, \tag{6.5}$$

where $\vec{R} = \vec{r} - \vec{r}'$. One can think of the current element $I\,d\vec{l}$ as a small piece of wire of directed length $d\vec{l}$ carrying current I. The net field $\vec{B}(\vec{r})$ due to the steady currents in a set of closed circuits is obtained by adding the contributions $d\vec{B}$ due to all of the current elements making up the circuits, i.e., by summing over circuits of closed path integrals around each circuit:

$$\vec{B}(\vec{r}) = \frac{\mu_o}{4\pi} \sum_j \oint_{jth\ circuit} \frac{I_j d\vec{l} \times \vec{R}}{R^3}, \tag{6.6}$$

This form of the Biot-Savart law can be easily rewritten to apply to a distributed current represented by current density $\vec{J}(\vec{r}')$ as follows:

$$\vec{B}(\vec{r}) = \frac{\mu_o}{4\pi} \int \frac{\vec{J}(\vec{r}') \times \vec{R}}{R^3} d\tau'. \tag{6.7}$$

The latter form of the Biot-Savart law is entirely equivalent to Maxwell's equations for the magnetostatic field $\vec{B}$.

6.2.2   The Magnetization $\vec{M}(\vec{r})$

When matter is subjected to an externally applied magnetic field, microscopic magnetic dipoles are induced in individual atoms. This happens in diamagnetic materials because magnetic forces on orbiting electrons induce an unbalanced current loop in the atom. In accord with Lenz's law, this induced current opposes change in the microscopic $\vec{B}$ field. In paramagnetic materials, the magnetic moments associated with the spins of unpaired electrons tend to align themselves with the applied field, and this reinforces the local $\vec{B}$ field. The diamagnetic effect is present in all materials, but it is usually overwhelmed by the paramagnetic effect when the atoms of the material have unpaired electrons.

In either the paramagnetic or the diamagnetic case, the induced time-averaged magnetic dipole moments in the individual atoms appear as a continuous distribution of magnetic dipole moment on distance scales much larger than atomic sizes. We represent this distribution of microscopic dipoles by a macroscopic magnetization field $\vec{M}(\vec{r})$, which is just the average dipole moment per unit volume at point $\vec{r}$.

Variations in $\vec{M}(\vec{r})$ over space, such as those that occur at material boundaries, are equivalent to unbalanced distributions of the induced microscopic currents in the atoms. These appear as macroscopic distributions of electric current associated with the motions of electrons that are "bound" to the atoms of the material. The density of bound current can be shown to be just the curl of the magnetization, i.e.,

$$\vec{J}_{bound} = \vec{\nabla} \times \vec{M}. \tag{6.8}$$

Bound current is the combined effect of many microscopic currents induced by the applied magnetic field, and is to be contrasted with the "free" current of charge carriers not bound to atoms (such as the free electrons in metals). The source of magnetic induction field $\vec{B}$ is the sum of both types of current:

$$\vec{J} = \vec{J}_{bound} + \vec{J}_{free} = \vec{\nabla} \times \vec{M} + \vec{J}_{free}. \tag{6.9}$$

6.2.3   The Magnetic Field $\vec{H}$

Bound currents both are produced by the applied field and are themselves sources of it. This leads to practical difficulties in the computation of fields. For example, one needs $\vec{J}_{bound}$ to compute $\vec{B}$ from the Biot-Savart law, but $\vec{J}_{bound}$ can't be known until $\vec{B}$ itself is known. For that reason, it is often convenient to use a new macroscopic magnetic field $\vec{H}$ defined as follows:

$$\vec{H} = \frac{\vec{B}}{\mu_o} - \vec{M}. \qquad (6.10)$$

In terms of this new field $\vec{H}$, the basic differential equations of magnetostatics become

$$\vec{\nabla} \cdot \vec{H} = -\vec{\nabla} \cdot \vec{M}, \qquad (6.11)$$

and

$$\vec{\nabla} \times \vec{H} = \vec{J}_{free}. \qquad (6.12)$$

This reformulation of the equations allows $\vec{H}$ to be computed from $\vec{J}_{free}$ alone if $\vec{\nabla} \cdot \vec{M} = 0$ or is negligible, because in that case the equations for $\vec{H}$ are essentially the same as those for $\vec{B}$ in the absence of magnetizable materials. Alternatively, if $\vec{J}_{free} = 0$ or is negligible and $\vec{M}$ is known, the equations for $\vec{H}$ become similar to the equations for $\vec{E}$ in electrostatics, but with density of electric charge replaced by the "magnetic charge density":

$$\rho_m = -\vec{\nabla} \cdot \vec{M}. \qquad (6.13)$$

When $\vec{M}$ changes abruptly, such as at a material boundary, it is possible to compute $\vec{H}$ in analogy with electrostatics by defining a "magnetic surface charge density":

$$\sigma = \hat{n} \cdot \vec{M}, \qquad (6.14)$$

where $\hat{n}$ is a unit vector normal to the material surface and directed away from it.

6.2.4   Magnetic Susceptibility and Permeability

The magnetization produced by the applied fields in diamagnetic and paramagnetic materials is "linear," which means that the magnetization is proportional to and parallel to the applied field, so that everywhere inside the material $\vec{B}$, $\vec{H}$, and $\vec{M}$ are parallel and proportional. In that case, it is useful to characterize the material through a magnetic susceptibility χ, which is defined to be the ratio of $\vec{M}$ to $\vec{H}$:

$$\vec{M} = \chi\vec{H}. \qquad (6.15)$$

The value of χ also determines the constant of proportionality between $\vec{B}$ and $\vec{H}$, which is called the permeability, μ:

$$\vec{B} = \mu_o(\vec{H} + \vec{M}) = \mu_o(1 + \chi)\vec{H} = \mu\vec{H} = \mu_r\mu_o\vec{H}. \qquad (6.16)$$

Since $\vec{M}$ and $\vec{H}$ have the same units (in this mks system), χ is dimensionless, as is the relative permeability μ_r.

6.2.5   Permanent Magnetization and Ferromagnetism

In "ferromagnetic" materials, such as iron and many alloys, there is a strong tendency for the spins of unpaired electrons in adjacent atoms to align in parallel. This results in a cooperative tendency for all atoms in regions of the solid material to develop parallel magnetic moments. However, the long-range interaction between magnetic dipoles makes it energetically unfavorable for very large regions to have parallel magnetization, so the material tends to subdivide itself into "magnetic domains," whose characteristic size represents a compromise between the short-range and long-range effects. Within each domain, the spins are parallel, but adjacent domains tend to have antiparallel spin orientations. Magnetic domains are submillimeter in size, yet many interatomic spacings in linear dimension.

The effect of applying an external magnetic field to a ferromagnet is to cause domains with magnetic moments parallel to the applied $\vec{B}$ to expand, and those with antiparallel magnetic moments to shrink, in rough proportion to the applied field. The net effect is cooperative paramagnetic behavior with very large effective magnetic susceptibility. When their domains expand or contract relatively freely in response to applied fields, materials are called "soft" ferromagnets. "Hard" ferromagnetic materials exhibit "permanent magnetization," which results from the phenomenon of "hysteresis." Hysteresis in ferromagnets is the dependence of the local magnetization field on the magnetic history of the material. Hysteresis happens when microscopic effects in the material strongly resist motions of domain "walls," so that the sizes of the domains cannot easily respond to changes in the external field.

The concept of magnetic susceptibility is not applicable to "hard" ferromagnets because the magnetization is no longer proportional to the field. On the contrary, a permanent magnet is one in which a strong magnetization exists in the material without any applied external field.

6.3 *Computational Considerations*

If the current $\vec{J}$ were known and specified at all points in space, computing $\vec{B}$, $\vec{H}$, and $\vec{M}$ at any given point would be a simple matter of integrating the Biot-Savart law. In the more realistic case of a magnetizable material in an external field, however, only the free currents responsible for the applied field are known. The bound currents induced in the material by application of the field can be computed only if the field is known, and vice versa. This is similar to the situation in electrostatics where the equilibrium configuration of induced bound charges in dielectric materials, or of free charges in conductors, both determines and is determined by the electric field. In electrostatics, the associated computational problems are dealt with through "relaxation methods," such as in the POISSON and GAUSS programs of an earlier chapter. We will use a somewhat different type of relaxation method here.

6.3.1 Field of a Permanent Magnet

In the case of a permanent magnet, there are no free currents, but there is a constant magnetization $\vec{M}$. Given this magnetization at all points in space, one can compute $\vec{H}$ in either of two ways: 1) find the bound current and and use it with the Biot-Savart law, or 2) find the magnetic charge and use it with the magnetic equivalent of Coulomb's law. In general, both the magnetic charge density $(-\vec{\nabla} \cdot \vec{M})$ and the bound current density $(\vec{\nabla} \times \vec{M})$ will be distributed throughout the permanently magnetized material, so that either method for computing $\vec{H}$ requires first differentiating $\vec{M}$, then integrating over the volume of the material. This general computational problem is not feasible to do with the hardware for which CUPS simulations are intended.

The simulation of permanent magnets in the MAGSTAT simulation is restricted to the special case where the magnetization is uniform within the magnet and directed parallel to an axis of cylindrical symmetry. This uniformity of the magnetization means that both the bound current and magnetic charge are confined to the surface of the magnet and specified once the magnetization and the shape of the surface are specified. (For similar reasons, the magnetic charge is also confined to the surface for the other special cases treated in MAGSTAT.) Given the bound current distribution, we can compute $\vec{H}$ from the Biot-Savart law.

Alternatively, given the magnetic charge distribution, we can compute the field from a magnetostatic equivalent of Coulomb's law. Only the units and associated multiplying constant make this computation different from the equivalent computation of electric fields from a static distribution of electric charge. Whether we use the Biot-Savart law or Coulomb's law, computing the field at any point in space requires doing an integral over the surface. Coulomb's law is somewhat simpler, so we use the magnetic charge in this simulation.

Numerical computation of a surface integral each time we want to find the magnetic field at some new point in space might be computationally expensive, but it can be done to an adequate approximation in the following way: We first imagine dividing the surface of revolution into narrow ribbon-like circular rings of magnetic charge each of width ds. At points sufficiently distant from the ribbon ($r \gg ds$), the field due to the ribbon is the field of a circle of charge with line density $\lambda = \sigma \, ds$. As an approximation, we replace the field of the ribbon by the field of a ring or circular line of charge. Then the field can be computed by summing the contributions from the rings.

6.3.2   The Field of a Ring of Charge

Imagine a circular ring of magnetic charge with radius R and line magnetic charge density λ centered on the origin and lying in the x-y plane. (To make the analogy with electrostatics clear, we will use the same symbols [q, λ, and σ] for magnetic charge distributions as are conventionally used in electrostatics for point, line and surface distributions of electric charge.) The field $\vec{H}(\vec{r})$ due to the ring can be computed in analogy with Coulomb's law by integrating over the ring:

$$\vec{H} = \left(\frac{1}{4\pi}\right) \int \frac{dq(\vec{r} - \vec{r}')}{|(\vec{r} - \vec{r}')|^3}, \tag{6.17}$$

where $\vec{r}'$ is the position of the differential magnetic charge dq. In cylindrical coordinates (ρ, ϕ, z), $dq = 2\pi\lambda R d\phi$. The field due to the ring has no axial component, and can be written

$$\vec{H}_{ring}(\rho, z) = H_\rho \hat{\rho} + H_z \hat{z}, \tag{6.18}$$

where

$$H_z(\rho, z) = \lambda R z I_0(\rho, z) \tag{6.19}$$

and

$$H_\rho(\rho, z) = \lambda R[\rho I_0(\rho, z) - R I_1(\rho, z)]. \tag{6.20}$$

Here $I_0(\rho, z)$ and $I_1(\rho, z)$ are integrals over ϕ:

$$I_n(\rho, z) = \int \frac{\cos^n(\phi) d\phi}{[z^2 + R^2 + \rho^2 - 2R\rho \cos(\phi)]^{3/2}} \tag{6.21}$$

The integrals $I_n(\rho, z)$ can be rewritten in a more convenient form as

$$I_n(\rho, z) = \zeta^{-3} \int \frac{\cos^n(\phi) d\phi}{[\zeta^2 + \alpha(1 - \cos(\phi))]^{3/2}} \tag{6.22}$$

where $\zeta = \sqrt{z^2 + (R - \rho)^2}$ is the distance to the nearest point on the ring, and α is the ratio $2R\rho/\zeta^2$. To evaluate $I_n(\rho, z)$ efficiently for large values of ζ we use an expansion of $\zeta^3 I_n(\rho, z)$ in Chebyschev polynomials.

6.3.3 Field Near a Magnetizable Material

Suppose we have a solid piece of uniform, linear, magnetic material in an applied field, so that $\vec{B} = \mu \vec{H}$, where μ is uniform except for the abrupt change at the interface between the material and the matter-free space around it. Also, suppose that the free current responsible for the applied field is zero except at very distant points in space. Then within the material and near it, we have

$$\vec{\nabla} \cdot \vec{H} = \rho_m, \tag{6.23}$$

and

$$\vec{\nabla} \times \vec{H} = 0, \tag{6.24}$$

where

$$\rho_m = -\vec{\nabla} \cdot \vec{M} = -\vec{\nabla} \cdot \left(\frac{\chi}{\mu} \vec{B}\right) = -\left(\vec{\nabla} \frac{\chi}{\mu}\right) \cdot \vec{B} - \frac{\chi}{\mu} \vec{\nabla} \cdot \vec{B}. \tag{6.25}$$

Since both χ and μ are uniform except at the the material boundary, and since $\vec{\nabla} \cdot \vec{B} = 0$ everywhere, the magnetic charge density ρ_m is zero everywhere except at the surface of the material. Since $\vec{M}$ is zero outside, the problem of finding $\vec{H}$ when the magnetization is known is again equivalent to finding the electric field when there is a known distribution of surface electric charge. The magnetic surface charge in this case is $\sigma = \hat{n} \cdot \vec{M}$, where $\hat{n}$ is a unit vector perpendicular to the surface and directed outward. Since the magnetization is proportional to $\vec{H}$, so is the magnetic surface charge

$$\sigma = \chi(\hat{n} \cdot \vec{H}_{in}), \tag{6.26}$$

where $\vec{H}_{in}$ is the field just inside the surface.

The surface charge distribution creates a discontinuity in the component of $\vec{H}$ perpendicular to the surface:

$$\hat{n} \cdot (\vec{H}_{out} - \vec{H}_{in}) = \sigma. \tag{6.27}$$

The magnetic charge depends on $\vec{H}$, but it is confined to the surface of the material, which considerably simplifies the problem of computing $\vec{H}$ at other points, and reduces the problem of finding self-consistent values of the vector $\vec{H}$ everywhere in the material to that of finding self-consistent values of the scalar σ at points on it's surface. This confinement of magnetic charge to the surface, plus its axial symmetry, thus reduces the three-dimensional problem to a one-dimensional one. Symmetry further reduces the problem to that of finding self-consistent magnetic charge distributions over only one-fourth of the pole-to-pole perimeter of the object.

6.4   Numerical Algorithms

In the computer code, we replace the distribution of magnetic charge over the solid surface by 50 discrete rings. Once the magnetic charge associated with each ring is known, the magnetic field $\vec{H}$ on any point in space can be computed by adding contributions from the rings, with each such contribution computed using Chebyschev polynomial approximants.

In the case of the field of a permanent magnet, the magnetization $\vec{M}$ is specified from the beginning, so that the magnetic charge $\sigma = \hat{n} \cdot \vec{M}$ is known and $\vec{H}$ can be computed directly. The case of a magnetizable material requires that the magnetic charge be computed first. The algorithm for that is as follows: Mirror symmetry about the x-y plane reduces the problem to finding the self-consistent magnetic charge on 25 rings. Let σ_i be the density of magnetic surface charge on ring i. Then Coulomb's law can be used to represent the field $\vec{H}_i$ just inside the surface of ring i as a sum over j of contributions from individual rings j, with each such contribution proportional to σ_j. Next, $\vec{H}_i$ can be used to compute σ_i through $\sigma = \chi(\hat{n} \cdot \vec{H}_{in})$. The net result is a set of 25 simultaneous linear equations of the form

$$\sigma_i = \chi(\hat{n}_i \cdot \vec{H}_o) + \chi \sum_{j}^{25} A_{ij}\sigma_j, \qquad (6.28)$$

where the coefficients A_{ij} depend on the relative positions and sizes of the rings i and j. Interested readers can consult the source code to see the details of how the A_{ij} are computed.

Once the coefficients A_{ij} are known, the equations are solved numerically. This could possibly be done by matrix inversion, but we instead use an iterative "over-relaxation" process that starts with a rough guess for the values of the σ_i. In each subsequent iteration, the new values of σ_i are modified by adding a fraction of the amount they changed in the iteration, after which the modified values are substituted in the equations to compute new values. The sum of the squared changes in the σ_i in each iteration is computed and used as a global measure of convergence. Once the relaxation iterations have converged on a set of unchanging values of σ_i, these values can be used to compute $\vec{H}(\vec{r})$ at any other point in space. $\vec{M}$ and $\vec{B}$ can then be computed easily.

6.5   About the Program

For each new simulation, the first step in the computation of a field is the determination of the magnetic charge on each of the 50 rings of charge used to approximate a continuous distribution over the surface of the solid. In the case of a permanently magnetized solid, this is done simply by finding the component of $\vec{M}$ perpendicular to the surface. In the alternative case of a magnetizable solid in an externally applied field, the charges on the rings are computed by the iterative over-relaxation method described above. Once the charges on the rings are found and stored, the magnetic field at any point in space can be computed by adding the contributions from the 50 rings and from the externally applied field.

6.5.1   Running the Program

The program begins with the sample computation and display of the magnetic field pattern ($\vec{H}$) for a set of default parameters, followed by superposition of a credits screen (which can be recalled at any time from the **FILE** menu). The press of any key or click of a mouse will activate hot key options and pull-down menus for analysis of the simulation currently displayed, or subsequent generation of user-generated field patterns.

An example of a field pattern generated by the MAGSTAT program is provided in Figure 6.1.

The first step in generating Figure 6.1's field pattern was to compute the components of $\vec{H}$ at each one of a square grid of points on the screen and to store them in matix form. Finding $\vec{H}$ at each point requires summing the contributions from 50 rings of magnetic charge, and the contribution from each such ring requires evaluation of a Chebyschev polynomial; this is a computationally intensive process, so generating the initial field display may be slow with some hardware. However, once $\vec{H}$ has been computed on the grid and stored in memory, subsequent displays of $\vec{H}$, $\vec{B}$, or $\vec{M}$ are generated much more quickly using the **CHOOSE** menu.

It is possible to speed up the computation by decreasing the dimension of the display grid from one of the options under the **DATA** menu. The penalty for increasing the speed with which a given field pattern is generated is a loss in fineness of the grid for the display of the fields. However, the fineness of the field grid does not affect the accuracy of the computation used to compute fields under the **TOOLS** menu items. If you wish a more detailed portrayal of the field under any of the three options under the **DISPLAY** menu at the expense of reduced speed of computation, you should choose a large grid size (i.e., a fine grid).

After the initial display of a field pattern, the program displays magnetic charge distribution at the lower left corner of the screen. This is done in the form of a plot of magnetic charge density versus distance along the surface. For all standard shapes allowed by the menu, this distance is measured along a semiperimeter of the cross section displayed on the screen, starting with the top polar point and ending at the bottom polar point. In the torus case, the plot is also surface charge density versus distance, but the distance is measured from the innermost point of the torus around and back to the same point. The part of the perimeter covered by the plot is shaded on the display in the same color as the plot. The **TOOLS** menu allows the user to redisplay this charge density plot at any time or to generate a similar display bound surface current, such as is plotted in Figure 6.1.

The **CHOOSE** menu allows the user to select any one of the three fields $\vec{H}$, $\vec{B}$, or $\vec{M}$ to display, and the **DISPLAY** menu offers three options for displaying them. The **Vector Grid** option displays the field by drawing colored arrows

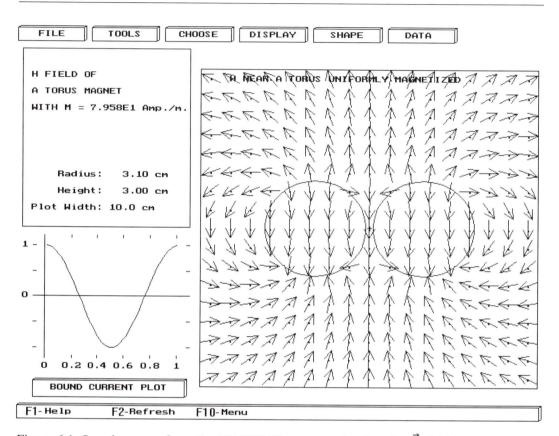

Figure 6.1: Sample screen from the MAGSTAT program showing the $\vec{H}$ field in and around a uniformly magnetized torus.

centered on each grid point. The arrows are drawn with uniform length, chosen to approximate the distance between nearest-neighbor points of the field grid, and with brighter colors for more intense fields (and vice versa). This display is meant to be reminiscent of an array of compass needles for plotting out field directions. Alternatively, the **Contours** option under **DISPLAY** draws a contour map of field intensity for the currently chosen field, and **3-D Surface** under **DISPLAY** illustrates the dependence of field intensity on location in the plane in yet a third way.

Once a field is displayed, the **TOOLS** menu provides several methods for examining it in quantitative detail. Magnitudes or directions of $\vec{H}$, $\vec{B}$, or $\vec{M}$ can be plotted versus distance along any mouse-selected line through the viewport, as illustrated in the sample screen shown in Figure 6.2.

Alternatively, the **Field Probe** option under **TOOLS** allows the user to use the mouse to explore the magnitudes and directions of $\vec{H}$, $\vec{B}$, and $\vec{M}$ at any point in the viewport. Use of these **TOOLS** options leaves additional lines and arrows on the field display, which you may want to eliminate by using the **Refresh** hot key, which simply redraws the current field pattern.

The **SHAPES** menu allows the user quickly to change the geometric shape of the magnetic solid, keeping the overall radius, aspect ratio, and other parameters exactly the same. This is useful for examining the effect of shape alone on the object. Both shape and size can be changed from the **DATA** menu.

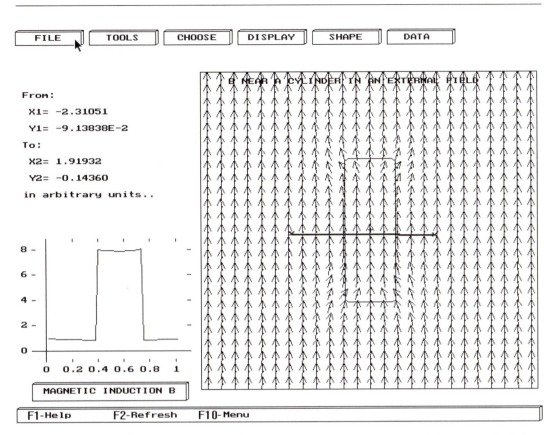

Figure 6.2: Sample MAGSTAT screen showing a plot of the magnitude of the magnetic induction field $\vec{B}$ along a line through the axis of a magnetizable cylinder in a uniform external field.

6.5.2   Approximations and Limitations of the Program

Tradeoffs must be made between computational speed and precision in almost any numerical computation. That is especially true in this simulation, because performing a meaningful, realistic, and interactive three-dimensional (3-D) computation of magnetic fields with reasonable computational speed pushes at limits imposed by CUPS-standard hardware. In particular, this means that the accuracy of computed values of fields in this simulation are not uniform across the displayed region.

The major difficulty with the computation is that the results tend to be inaccurate at points very near the surface of the solid material being simulated. This effect is due almost entirely to the approximation in which a continuous distribution of surface magnetic charge is replaced by a set of discrete rings of charge. When the nearest distance from a field point to any ring becomes comparable to the distance between rings, the sum of contributions from the rings is no longer a reasonable approximation to an integral over a continuous distribution of surface charge.

The user may become aware of this inaccuracy when the field near the surface occasionally points in strange, unphysical directions near the surface. When that happens, the magnitude of the field quite near the surface will also be inaccurate; this can reveal itself as strange spikes in 3-D plots or as odd blips near the surface in contour plots. This inaccuracy decreases rapidly with distance from the surface, and has no additional effects on simulations of the field of a permanent magnet. However, there can be secondary inaccuracies in the results of the relaxation calculations for large values of magnetic permittivity and near sharp corners in the surface of some solids.

6.6   Exercises

6.1   Magnetic Dipole Field

Generate the magnetic field pattern of a permanent magnet for a variety of shapes and aspect ratios (i.e., ratio of length to width) and observe the dependence of the field on shape and size; then do the following exercises:

a. By using the **Contour** option under **DISPLAY**, verify that the field ($\vec{B}$ or $\vec{H}$) far away from the magnet (distance greater than the linear size of the magnet) has the same general pattern independent of the shape of the magnet, but not so for points nearer the surface. Using the **Field Probe** option under **TOOLS**, measure and record the angular direction of the field in 30-degree intervals around a sphere at a distance from the center of 1.1 times the radius, and again at a distance 3.0 times the radius. Repeat for the case of a double cone with a square cross section (aspect ratio = 1.00). Plot field direction versus angular position as "four curves" four cases on the same piece of graph paper. Explain similarities and differences between the four different curves. Compare your plots with theoretical predictions for a dipole field.

b. Use **TOOLS/Field Probe** to measure the magnitude of $\vec{H}$ versus distance along some line through the origin at the center of the viewport. Plot a logarithm of field magnitude versus logarithm of distance. Your plot should become a straight line at larger distances. Measure the slope of this straight line and compare it with what you would expect for a dipole field.

6.2   Dependence of the Field of a Permanent Magnet on Size

Choose three permanent magnets of unit aspect ratio (length = width) and equal linear dimension, but with different shapes. Use formulae for the volume of a sphere, cylinder, and cone and the **Field Probe** option under **TOOLS** to verify that the magnetic dipole field produced by a permanent magnet is proportional to its net dipole moment, i.e., $\vec{M}$ times volume. (One way to do this is measure the magnitude of the field at a few standard locations a sufficient distance away from the object.)

6.3   Dependence of the Field of a Permanent Magnet on Shape

Answer the following questions by generating the field external to permanent magnets of different shapes and aspect ratios:

a. Use **TOOLS/Field Probe** to measure and compare the field near the poles of a spherical magnet with the field near the "equator" measured the same way. Use a handbook to look up the values needed to make the same comparison for the magnetic field of the Earth. Are the results the same, within expected errors? Why or why not?

b. How does the field near the poles of an ellipsoidal magnet compare with the field near its equator? Plot the ratio of the field at the poles of an ellipsoidal permanent magnet to the field at its equator as a function of aspect ratio (length/width). For a constant mass and magnetization, would a prolate or oblate permanent magnet generate a stronger external magnetic field?

c. Where around the surface of a cylindrical permanent magnet is the external field the strongest and where is it weakest? Does the answer depend on the aspect ratio (length/width)?

d. It is observed that iron filings tend to congregate at the sharp edges of permanent magnets. Make a hypothesis as to why this is so, then verify your hypothesis using the simulation.

6.4   Induced Magnetization

When magnetization is induced in a linear medium by placing a solid in an external field, would you expect the magnetization to be proportional to the magnitude of the external field? Would you expect the magnetization induced at the center of such an object to be independent of the shape of the object? Would you expect it to depend on the ratio of length to width? Would you expect it to be proportional to permittivity μ or to susceptibility χ? Make a hypotheses about how the magnetization at the center of an object depends on these factors, then use **Field Probe** under **TOOLS** to test your hypotheses. In each case, write your hypothesis explicitly, describe the procedure for testing it, record appropriate data, and state whether or not the hypothesis is confirmed.

6.5   Boundary Conditions on $\vec{B}$ and $\vec{H}$

The requirement that $\vec{\nabla} \cdot \vec{B} = 0$ implies that the component of $\vec{B}$ perpendicular to the surface of a magnetizable material must be continuous across the surface. Similarly, the requirement that $\vec{\nabla} \times \vec{H} = \vec{J}_{free}$ implies that the component of $\vec{H}$ parallel to such a surface must be continuous across it when there is no free current.

a. Generate the field of a magnetizable sphere in a uniform external field and use the **Plot What** option to show that the magnitude of $\vec{B}$ is continuous across its surface when $\vec{B}$ is perpendicular to the surface, but not otherwise.

b. Similarly, use the **Plot What** option for a magnetizable cylinder to show that the magnitude of $\vec{H}$ is continuous across the surface only when $\vec{H}$ is parallel to it.

c. Generate the field pattern near a magnetizable double-cone-shape solid and use the **Explore** option to see how the direction of the field changes at the surface. Do this for the two cases $\mu = 5$ and $\mu = 0.5$. Note the difference between the two cases and give a qualitative explanation.

d. Use the **Explore** option to measure the magnitude and direction of $\vec{H}$ at five points lying on opposite sides of the above double-cone surface when $\mu = 5$. Compute the components of $\vec{H}$ parallel to the surface to verify that it is the same on either side. Repeat for the perpendicular component of $\vec{B}$.

e. Derive and use the simulation to verify a formula relating the directions of $\vec{H}$ and $\vec{B}$ on either side of a boundary between a linear material of susceptibility μ and free space.

6.6   Magnetic Charge Distribution

The surface distribution of magnetic charge, $\hat{n} \cdot \vec{M}$, is plotted each time a new field pattern is generated. In the case of a permanent magnet with uniform $\vec{M}$, this distribution is determined by the shape of the magnet in a straightforward way, but the magnetic charge distribution when the magnetization is induced by an external field can difficult to predict.

a. It can be proved analytically that an ellipsoid in a uniform applied field will develop a uniform magnetization, so that the magnetic charge should be distributed similarly in this simulation for both a permanently magnetized ellipsoid (or sphere) and an ellipsoid (or sphere) in an external field. Verify that this is true.

b. Changing only the shape, from ellipsoid to cylinder to double cone, note how the distribution of magnetic charge over the surface changes. Repeat this for both a paramagnetic material and a diamagnetic material $\mu = 5.0$ and $\mu = 0.20$.

6.7   Bound Magnetic Current

The magnetic fields in the presence of magnetized matter can be computed from either the bound currents (using the Biot-Savart law) or the magnetic charge (using the magnetic equivalent of Coulomb's law, as we actually do it). In the examples treated in this simulation, both the magnetic currents and the magnetic charges are distributed over the surface. A menu item allows you to select either bound current or magnetic charge for plotting. Examine the distribution of both bound current and magnetic charge over the surface for the case of permanently magnetized cylinder (bar or disk magnet) with uniform $\vec{M}$; notice that the distributions of magnetic charge and magnetic currents do not overlap. Determine whether there are any shapes or aspect ratios where the magnetic charge distribution looks similar to the magnetic current distribution.

6.8   Creating New Shapes

Examine the Pascal code that sets up each of the four different types of surfaces. Notice that all shapes share a set of symmetries, such as mirror symmetry about the x-y plane and axial symmetry around the z-axis. The perimeter of the surface must be broken into 100 points corresponding to 50 equally spaced rings. Note that the local unit normals to the surface must be specified, too. Modify the code to add additional shapes having those same symmetries, such as the following:

1. A solid of rotation whose cross section is a regular hexagon with a flat "pole."

2. A solid of rotation whose cross section is a regular hexagon with a vertex at the pole.

3. A cylinder with concave spherical surfaces (rather than a plane surface) at the top and bottom.

4. A cylinder with a concave cone (rather than a plane surface) at the top and bottom.

5. A ring with a square cross section (a torus has a circular cross section).

6.9 Superconductors as Perfect Diamagnets

When a type-I superconductor is placed in an external magnetic field it develops a thin layer of free current near the surface that creates an equal and opposite magnetic field inside the solid. The result is that the superconductor becomes impermeable to externally applied fields. This ability of a superconductor to expel magnetic fields is called the Meissner effect. The Meissner effect is similar to perfect diamagnetism, which can be treated in this simulation by setting the permeability μ equal to zero. Measure $\vec{H}$, $\vec{B}$, and $\vec{M}$ at the center of the sphere for a constant applied field and values of μ between 2.0 and 0.0. Plot all the z-components as functions of μ to find all three fields inside a perfect diamagnet. Compare the results with what you would expect if the interior fields were entirely due to free electric current on the surface (the Meissner effect case).

6.10 Magnetic Field Near a Perfect Diamagnet

Generate the $\vec{B}$ field for $\mu = 0$ around a torus with an aspect ratio that makes the radius of the central hole through the torus (the donut hole) very small (0.48, for example). Where does $\vec{B}$ have its greatest intensity in this case? Repeat for several other shapes with the same size and aspect ratio. Formulate an explanation of your results in terms of continuity of magnetic field lines, remembering that lines of $\vec{B}$ have no sources or sinks, and that the intensity of the field is proportional to the density of field lines.

6.11 Magnetic Flux Through a Diamagnetic Torus

Generate the $\vec{B}$ field for $\mu = 0$ around a torus for five different aspect ratios between 0.49 and 0.20, keeping the overall radius and applied field the same. For each case, estimate the total magnetic flux through the central hole in the torus by assuming that the magnetic field is uniform in that region (Can you think of a better approximation?). Compare that flux with the total flux intercepted by the torus, i.e., the value of the applied field multiplied by the total area of the torus (πr^2). Are the results about what you would expect if the torus were a perfect superconductor? Explain.

6.12   **Designing a Permanent Magnet**

Large electromagnets can produce uniform external fields of a few tesla (a few tens of kilogauss). A permanent magnet is to be made by placing a piece of ferromagnetic alloy in this external field.

a. Suppose the piece of alloy has permeability $\mu = 10$ and is placed in an applied magnetic field $\vec{H}_o$ of 1 tesla. Use the simulation to determine the shape and aspect ratio that produces the maximum magnetization at its center under these conditions. Does the result depend strongly on shape (i.e., cylinder vs. ellipsoid) or aspect ratio or both? Repeat for $\mu = 5$ and $\mu = 2$.

b. Assume that the maximum magnetization achieved above remains as a uniform permanent magnetization after the external magnetic field is turned off. The permanent magnet created this way then has an external field $\vec{H}$. Use the simulation to determine the place where this external field is greatest and compare its value there with the original 1 tesla applied field. Repeat for the $\mu = 5$ and $\mu = 2$. Is this external field proportional to μ? Is it proportional to χ? Is it proportional to $\vec{H}_o$?

c. Is it possible by this hypothetical method to make a permanent magnet that has a magnetic field much larger than the magnetic field ($\vec{H}_o$) that created it? If so, predict how large a field could be produced with, say, $\mu = 1000$ and $H_o = 1$ tesla.

Bibliography

1. Lorrain, P., Corson, D. P., and Lorrain, F. *Electromagnetic Fields and Waves.* San Francisco: Freeman Press, 1988.

2. Jackson, J. D. *Classical Electrodynamics.* Chapter 5: Magnetostatics. New York: John Wiley and Sons, 1975.

3. Wangness, R. K. *Electromagnetic Fields.* New York: John Wiley and Sons, 1979.

7

Field of a Moving Charge

Ronald Stoner

> All my Flatland friends—when I talk to them about the unrecognized dimension that is somehow visible in a Line—say, "Ah, you mean brightness"; and when I reply, "No, I mean a real dimension," they at once retort, "Then measure it, or tell us in which direction it extends"; and this silences me ...

> —Edwin A. Abbot in *Flatland*.

7.1 Introduction

From a fundamental, microscopic point of view, electromagnetic fields originate from elementary particles that carry electric charge. It is the acceleration of these charged elementary particles that is the ultimate source of electromagnetic radiation. This simulation provides a method for investigating the relationship between the motion of one such charged particle and the electromagnetic fields it produces.

We live in three-dimensional space, but the visualization of electromagnetic fields in three dimensions presents a daunting challenge to the display technology of today's personal computers. Happily, though, nearly all of the important physics of the production of the radiation field can be investigated by examining the fields in the planes of motion that are produced by a particle moving in those planes. Therefore, although the electromagnetic fields treated in this simulation are three-dimensional, they are displayed only in the plane of motion of the charge. Because all vector fields in this special case of two-dimensional motion are directed either parallel to or perpendicular to that plane, their directions can be displayed in a manner common to textbook illustrations: Fields parallel to the plane are represented by vector arrows, fields directed into the plane are represented by plus signs, and those directed outward are represented by dots.

7.2   The Electric Field

At all points in the plane of motion, the electric field $\vec{E}$ of a charge moving in two dimensions is parallel to the plane. What's more, the divergence of this electric field, $\vec{\nabla} \cdot \vec{E}$, is zero except at the position of the charge itself. These two characteristics allow $\vec{E}$ to be represented, at least in this plane of motion, by a two-dimensional plot of continuous field lines originating on the charge.

As will be explained below, there are computationally fast, accurate algorithms for drawing the electric field lines around a moving point charge. For that reason, and for the other reasons explained above, this simulation concentrates on the behavior of the electric field of a moving particle. The moving charge in the simulation is assumed positive, so the electric fields are, by the same assumption, directed away from it.

Coulomb's law implies that electric field lines can originate or terminate only on electric charge. A given line of electric field in space is continuous and unbroken from its origin on some positive electric charge to its eventual termination on a negative electric charge. When an electrically charged particle moves through space, those electric field lines that originate (or terminate) on it must respond by moving, stretching, and bending as if they were strings tied firmly to it.

The resulting behavior of the electric field of a moving point charge can be complex, but this behavior must conform to fundamental laws of physics that are quite simple. In fact, the behavior of the electric field follows directly from Gauss' law and the Lorentz transformation of special relativity. The mathematical simplicity of the underlying laws, combined with the complexity of the resulting field-line patterns, make the electric field of a moving charge an ideal subject for computer simulation and visualization.

7.2.1   A Point Charge in Uniform Motion

Coulomb's law describes the electric field of a point charge at rest as having an isotropic radial pattern centered on the charge. This is the field pattern observed in the "rest frame" of an unaccelerated charge. An inertial observer in a different frame of reference (e.g., the "laboratory frame") will also observe a radial field-line pattern with the point charge at its center, but both the charge and the electric field pattern will be in motion with the same constant velocity. The Lorentz invariance of electric charge, when combined with Gauss' law, guarantees that the total number of field lines leaving the charge is the same in both reference frames, but the spacing in angle won't be isotropic when the charge is in motion.

Since the electric field is known in the rest frame, the field-line pattern in any other inertial frame can be inferred by simple Lorentz transform. The result is that the electric field pattern is Lorentz contracted in the direction of motion; this means the whole field pattern is compressed along that direction. This contraction of the field line pattern diminishes the field-line intensity in directions along the straight trajectory of the charge and intensifies it in the perpendicular direction.

The above qualitative description of the field of a moving charge can easily be translated into a prescription for computing the directions of field lines radiating from the moving charge. In the rest frame, the isotropic nature of the field around the charge means that the angles between adjacent radial field lines are all the same. If Θ_{rest} is the angle a field line makes with the direction of motion in the rest frame, then the angular direction of the same field line in the laboratory frame is given by

$$\Theta_{lab} = \arctan(\gamma \tan(\Theta_{rest})), \qquad (7.1)$$

where γ is the Lorentz factor $1/\sqrt{1 - (v/c)^2}$.

This Lorentz transformation of the field-line pattern into the laboratory frame decreases the angular distance between field lines that are perpendicular to the velocity, and increases the angular distance between field lines that are roughly parallel to the velocity. This effect is the same as if the space between the lines were Lorentz contracted, but is actually a result of the way the two components of $\vec{E}$ transform as components of a four-tensor under a Lorentz transformation.[7] Since the magnitude of the electric field is proportional to the density of field lines, the effect is to increase the field intensity in the direction perpendicular to the motion and to reduce it in the forward or backward directions. At velocities near the speed of light, where the Lorentz factor is large, the isotropic pattern of the rest frame becomes flattened along the direction of motion into a disk-like bundle that moves along with the particle.

7.2.2   An Accelerated Charge

A good elementary description of how acceleration of a charged particle modifies its electric field is given in standard texts, such as Purcell.[1] A brief explanation is given below, along with a introduction to how the process can be simulated numerically.

Suppose a point charge in uniform motion suddenly changes its velocity, in either magnitude or direction, and subsequently moves with this new constant velocity. The electric field at some distance from the charge cannot be affected immediately, because information about the motion of the charge travels at the finite speed of light. Therefore, the field pattern far away will continue to behave as if the velocity of the charge had not changed. That is, the lines of $\vec{E}$ will continue to move at the original velocity and to converge toward where the point charge would have been had there been no velocity change. The effect on the field pattern is illustrated in the Figure 7.1, which is a sample screen from the ACCELQ program showing the field produced by a charge in linear motion that suddenly decelerates to rest.

Information on the changed motion of the charge will reach points nearer to it more quickly. This means that the field lines near the charge will quickly begin to reflect the new velocity of the charge and to move with it at that new velocity, which is zero in Figure 7.1. The "near field" close to the charge and the "far field" (which is still "unaware" of the acceleration) will be moving at two different velocities, and will diverge radially from two different points in space. The degree and direction of the Lorentz contraction effect will generally be different, too.

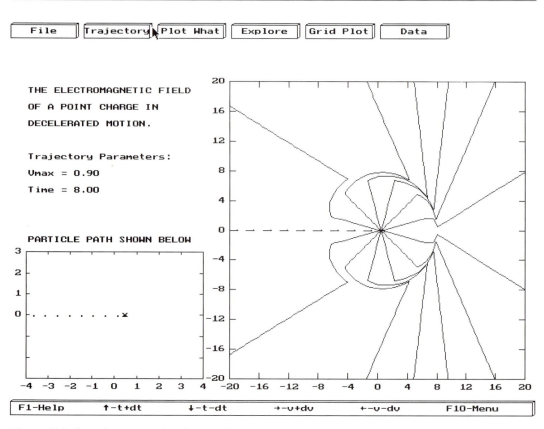

Figure 7.1: Sample screen showing the field generated when a fast-moving charged particle decelerates to zero velocity.

Between the near field and the far field, there will be a transition region that expands at the speed of light away from the point of the acceleration; the field in this transition region carries information about the acceleration that created it, and it will have non-radial components. The radial thickness of the transition region will be the product of the speed of light and the time required for the velocity change. The greater the acceleration for a given velocity change, the thinner the transition region, the more closely packed the field lines, and the greater the field intensity.

Electric field lines can neither terminate nor intersect except where there is electric charge, so the lines near the point charge and those far away must retain their continuity. Thus the field lines must bend and stretch in the expanding transition region so as to remain connected with both the lines of the near field and those of the far field. If the acceleration and speed of the particle is large, the field lines in the transition region will generally be relatively dense, and so the electric field intensity there will be proportionately larger.

As the transition region sweeps by a stationary observer in the laboratory frame (i.e., the reference frame of the computer screen), it will appear as a pulse of strong field carrying energy with it. Thus the expanding transition region represents a pulse of electromagnetic radiation produced by the acceleration of the charge.

7.3 Retarded Fields and Potentials

The discussion so far has described only the electric field $\vec{E}(\vec{r}, t)$. A moving charge also produces a magnetic field, $\vec{B}(\vec{r}, t)$, a scalar potential $\Phi(\vec{r}, t)$, a vector potential $\vec{A}(\vec{r}, t)$, and an energy flow field, or Poynting vector $\vec{S}(\vec{r}, t)$. The ACCELQ program allows all of these fields to be computed at arbitrary points and plotted along arbitrary lines through the plane of motion.

The scalar and vector potentials, $\Phi(\vec{r}, t)$ and $\vec{A}(\vec{r}, t)$, resulting from the motion of a point charge q can easily be computed from Lienard-Weichert formulas:

$$\Phi(\vec{r}, t) = \frac{1}{4\pi\epsilon_o} q[1/\kappa R]_{ret} \tag{7.2}$$

and

$$\vec{A}(\vec{r}, t) = \frac{\mu_o}{4\pi} q[\vec{v}/c\kappa R]_{ret}, \tag{7.3}$$

where $\vec{R}$ is the displacement between the point $\vec{r}$ where the fields are to be computed and the position $\vec{r}_q$ of the charge. The subscript "*ret*" in the above equations means that the quantities in square brackets are to be computed at the "retarded time," which is the time when the wavefront arriving at point $\vec{r}$ at time t was emitted, i.e., $t - \frac{R}{c}$. The quantity κ, which also must be evaluated at the retarded time, is defined to be

$$\kappa = 1 - \vec{v} \cdot \vec{R}/cR. \tag{7.4}$$

Because Φ and $\vec{A}$ must be evaluated at the retarded time, the formulas for $\vec{A}$ and Φ are implicit; that is, they can be computed only after the retarded time itself is computed, and the retarded time itself is not known explicitly. As will be discussed below, the simulation computes the retarded time, when needed, by numerical solution of a transcendental equation representing the requirement that the distance between $\vec{r}$ and the position $\vec{r}_q$ at the retarded time be the speed of light times the difference between viewing time, t, and the retarded time.

Formulas for the electric and magnetic fields $\vec{E}$ and $\vec{B}$ can be derived by careful differentiation of the retarded potentials $\vec{A}$ and Φ. These also must be evaluated at the retarded time. Such formulas for $\vec{E}$ and $\vec{B}$ from a moving point charge can be found in standard textbooks,[1,7,8] and are part of the Pascal code for this simulation.

7.4 Computational Algorithms

The numerical algorithms used in the simulation are discussed in this section.

7.4.1 Geometrical Construction of Field Lines

The velocity of an accelerated particle changes continuously, but the velocity is definite at each instant of time. It is useful to think of an accelerated charged particle as producing, at each instant, a spherically expanding signal, or wavefront,

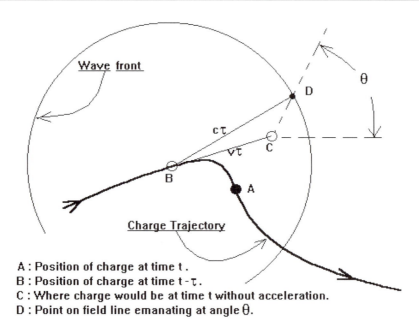

A : Position of charge at time t .
B : Position of charge at time t - τ.
C : Where charge would be at time t without acceleration.
D : Point on field line emanating at angle θ.

Figure 7.2: Geometrical construction for finding a point on a line of $\vec{E}$.

which carries information about the instantaneous velocity at that instant. As it expands at the speed of light, the spherical wavefront remains centered on the position of the particle at that same instant it was emitted (which is the "retarded time" described in the above section on retarded potentials and fields).

Each such spherical wavefront carries with it information on the directions of field lines emanating from the charge at the time the sphere began its expansion. These directions for each such sphere can be inferred from a relatively simple geometric construction as follows:

1. Draw a circle of radius $c\tau$ centered on the position of the particle at the time $t - \tau$, where t is the time of viewing in the observer's reference frame. The particle position at $t - \tau$ is point B in Figure 7.2; the circle is the intersection of the expanding wavefront with the plane of motion at the later time t, when the particle has moved on to point A in the figure;

2. Find the position C the particle would have at time t if it had continued without acceleration by drawing a line segment of length $v\,\tau$ from the center of the circle (i.e., point B) in the direction the charge was moving at time $t - \tau$.

3. Find the direction of a given field line by applying the Lorentz contraction formula to find the angle θ in the figure. Draw a straight line in that direction from point C. The intersection of this line with the circle is a point, D, on that particular field line at time t.

4. Repeat step 3 for every field line to find the intersection of every line with the sphere.

The resulting set of points on the circle locate the position of each field line intersecting that particular wavefront at time t. Since none of this construction depends on what happens to the particle after time $t - \tau$, the directions to these field-line points from point B are independent of the particular time t used to make the construction. That is, the set of directions is fixed for each wavefront at the time the wavefront is created; the directions are determined by the velocity of the charge at that point in time, and they are carried with the wavefront as it expands at the speed of light.

Constructing the pattern of electric-field lines is a matter of connecting corresponding points, one point on each circle, that correspond to the same field line. For a moving charge, each such wavefront will have a different center, and the angular spacing of the field points on each will also generally be different, leading to the stretching and bending of the field lines.

The above geometric construction is the basis for generating all plots of electric field patterns in this computer simulation, but two different algorithms are employed for the different cases of real-time interactive animation (the QANIMATE program) and "snapshots" (the ACCELQ program). Both algorithms use the retarded time itself as an independent variable, thereby avoiding the computationally expensive problem of computing a retarded time for each point on a field line. Both algorithms are described below.

7.4.2 Interactive Animation of Field Lines

It is helpful in understanding the algorithm used in the QANIMATE program to think of a point on a field line metaphorically, like a person who is "born" as one "generation" of a "family," has an active life, "dies," and is "buried." The points on a given field line represent the members of a family, one individual per generation. All points in a given generation are born simultaneously at the charge. The line connecting all points in a given generation is a wavefront circle, centered on the position of the charge when the front was created. The center of a wavefront does not change, but its radius increases at the speed of light. The direction from the center of the circle to a point is where a given field line meets the wavefront, and does not change as the wavefront expands.

In the charge's rest frame, the angle a field line makes with the horizontal is an integral multiple of 360 degrees divided by the number of points in a generation (i.e., the number of field lines shown), which is a constant number set by the user from a data screen. Since the wavefront is traveling at light speed, no information about the charge's subsequent motions can affect the direction in which the point moves with the expanding wavefront. So, a point on a given field line is always moving radially from the place where the charge was when the wavefront it is on was created, and in a direction determined at that place and time, i.e., the point's birth.

The charge travels in a path determined by keyboard control. The charge's speed may never exceed the speed of light (a velocity of 1.0 in the units used here). At each increment of time, the current acceleration and velocity are used to update the velocity and position of the charge, which in turn determine the directions for all the points in the new generation.

An approximation to a given field line is drawn by connecting all of the points in the same "family" in different generations. Once "born," points continue to move until they leave the viewing area, or "die." Dead points are only kept track of for as long as they are needed for determining the end points of displayed segments of the field lines; then they are deleted from memory ("buried"). A dead point is buried when it has no living neighbors (children or parents) in its field line (family).

Object-oriented programming (OOP) is used extensively in the QANIMATE program. The points themselves are objects; these objects are packaged with pointers to make records; lists of these records are used in larger objects. The **CircleList** is a large object, as is each **FieldLine**. The **FieldLines** then make up another linked list—the **FieldLinePtr** structure.

An animation technique called "page flipping" simulates the development of a field pattern in real time. In this technique, the field pattern at one viewing time is computed and drawn on a hidden page while a visible page is displayed showing a previously computed pattern at an earlier time. The pages are then "flipped," and the newly drawn pattern is displayed while a new animation frame is drawn on the newly hidden page.

7.4.3   Algorithm for Plotting Smooth Field Lines

Computation speed becomes less important when producing a "snapshot" of the field pattern, where producing a smooth, accurate representation of each line becomes more important. Therefore, the algorithm used in the ACCELQ program is a variation of one described by Hamilton and Schwartz.[3]

In this algorithm, each field line is drawn individually, starting at its origin on the electric charge. The trajectory of the particle in the plane of the screen is stored as an algorithm for computing the two Cartesian coordinates of position, velocity, and acceleration at any time. Each field line is traced outward from the position of the charge by stepping backward in time, i.e., incrementing the value of τ, and stepping from one wave front outward to one created at an earlier time, $t - \tau$, in the particle's motion. At each such time in the particle's previous motion, the computational equivalent of the geometric construction described above is done to trace the field line from one wave front to the next.

The "smoothness" with which a given field line is reproduced by this procedure is determined by the length of the straight-line segments drawn between successive field points. Very short line segments produce a very smoothly curved field line at the expense of increased computation time, and vice versa. The simulation allows the user to change a parameter, SegL, which adjusts the maximum length of such a line segment. If the length of the line segment between two such field points exceeds this maximum length, the time step is decreased; likewise, if the segment is shorter than half this maximum length, the time step is increased so as to maintain roughly constant smoothness.

Once a snapshot of the electric field pattern is produced, the simulation allows various other fields to be computed at selected points in the plane of the screen (the **EXPLORE** menu option), or along mouse-selected cuts through the plane of the screen (the **PLOT WHAT** menu option). This requires still a third algorithm, as follows:

To compute the fields due to a moving charge at a given point in space at a given point in time, we need to know the location and velocity of the charge at a precise previous time—the "retarded time" when the wave front passing the given point at the given time was created. This can be done by finding at what time in the past the charge was at a distance from the point equal to the distance light can travel in the interim. That is, it means solving the transcendental equation

$$|\vec{r}_o - \vec{r}(t_o - \tau)| = c\tau \tag{7.5}$$

where t_o and $\vec{r}_o$ are the time and position where the field is desired, $\vec{r}(t)$ is the formula for the position of the charge as a function of time (i.e., the trajectory), and τ is the light travel time.

The above equation is solved numerically for the value of τ, with the result that the position, velocity, and acceleration of the charge can be computed at the retarded time $t_o - \tau$. This allows the fields to be computed from explicit Lienard-Weichert formulas. Such explicit formulas can be found for the electric field, the magnetic field, the scalar and vector potentials, and the Poynting vector field in standard textbooks.[1,7,8]

7.5   *Using the Simulation*

The simulation comes in two distinct parts—QANIMATE, an animation part where the user can interactively accelerate and guide a charged particle, and ACCELQ, a still-frame (or "snapshot") generator with predefined particle trajectories. The animated part has some of the character of a video game, and the user may require a short practice period to gain dexterity in guiding the particle.

7.5.1   ACCELQ—Generating Still-Frame Field Patterns

As was discussed in the algorithms section above, the computational requirements of animation—in particular, the need for computational speed—place burdens on machine resources that can limit other options. For that reason, ACCELQ provides a more robust range of options for investigating electromagnetic fields than QANIMATE, and more accurate, smoother depiction of electric field lines as well.

ACCELQ allows the user to choose from among a variety of predefined trajectories. The trajectories are arranged in a menu in rough order of increasing complexity, but each is characterized by an adjustable parameter, v_{max}, which is the maximum speed attained by the particle in that trajectory (in units c = speed of light = 1). The user is allowed to set both v_{max} and a viewing time t from a data screen, and to increment or decrement them with "hot keys." Each time a new trajectory is selected from the menu, the motion of the charge in that trajectory is shown in animation at the lower left of the screen and the electric field pattern for the current values of v_{max} and t is plotted in the main viewport. An example of such a screen is shown in Figure 7.3.

Any change in v_{max} or t through the hot keys or data input screen results in a new plot of the field-line pattern. This feature allows a crude type of animation through simple repeated pressing of a hot key to draw a sequence of frames at equally spaced intervals of viewing time. The effect of varying the magnitude of

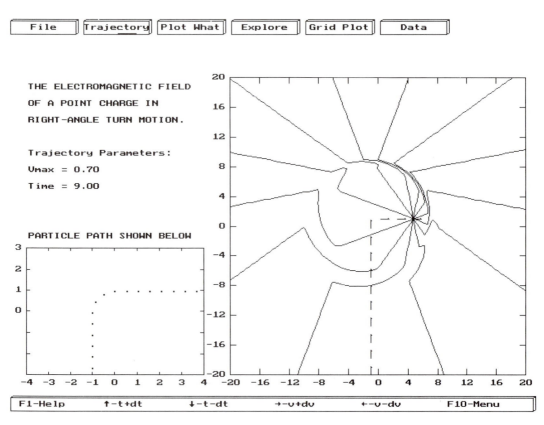

Figure 7.3: Sample screen showing the field generated when a charged particle makes a right-angle turn.

the acceleration can be animated in a similar way by pressing the hot key that increments v_{max} produced by a moving charge.

7.5.2   Exploring and Graphing the Fields

Once an electric field pattern is generated on the screen, the ACCELQ simulation allows several options for investigating it and the other electromagnetic fields. The **GRID** menu option causes each of the vector fields $\vec{B}$, $\vec{A}$, and $\vec{S}$ to be plotted at each point on a uniform grid to reveal the pattern of each. The **EXPLORE** menu option allows the mouse to be used as a field probe.

In the **PLOT WHAT** menu option, the mouse can be used to define a line through the plane; any one of the fields can be graphed as a function of position along such a user-selected line or as a function of time. This latter option is especially useful in investigating the "shape" of the radiation pulse produced by the accelerating charge, and how this pulse shape changes with the magnitude of v_{max}. Because the predefined trajectories involve accelerations over fixed intervals of time or distance, changes in v_{max} also represent changes in acceleration.

7.5.3   Predefined Trajectories

The still-frame portion of the simulation employs trajectories defined in a Pascal procedure. For each such trajectory, the Pascal code contains a prescription for computing the components of position, velocity, and acceleration as explicit functions of viewing time. Each such trajectory has the same adjustable parameter, v_{max}, which is the maximum speed (in units $c = 1$) attained by the particle during its motion in that trajectory. In the two cases where the only acceleration is centripetal, v_{max} represents the constant speed of the particle.

Discontinuities in velocity are avoided in all pre-defined trajectories, both because infinite acceleration is unphysical and because such discontinuities would interfere with algorithms for computing the retardation time. The different predefined trajectories are briefly described below:

LINEAR ACCELERATION occurs during the time interval $0 < t < 1$, during which the speed increases from $v = 0$ to $v = v_{max}$; so v_{max} is the value of the acceleration. For times $t < 0$ the charge is at rest at the origin.

LINEAR DECELERATION at constant acceleration $-v_{max}$ also occurs during $0 < t < 1$, bringing the charge from a speed of v_{max} to rest at $t = 1$.

SIMPLE HARMONIC MOTION is periodic with angular frequency $\omega = v_{max}$ and unit amplitude, so that the acceleration also varies harmonically with amplitude $\omega^2 = v_{max}^2$.

CIRCULAR MOTION is counterclockwise with uniform speed of v_{max}, and with unit radius.

The RIGHT-ANGLE TURN is by clockwise quarter-circle arc of unit radius, starting at time $t = 0$, and changing the direction of motion from screen bottom-to-top to screen left-to-right. The speed is constant at v_{max} throughout.

The SINUOUS trajectory has uniform velocity in the vertical direction and simple harmonic motion with unit amplitude in the horizontal direction. These two components of velocity are equal at the midpoint of the simple harmonic motion, where the speed is v_{max}.

The FIGURE-8 trajectory is a superposition of vertical and horizontal motions that are both simple harmonic, but at different frequencies. The vertical frequency is double the horizontal frequency.

7.5.4   QANIMATE—Steering the Charge

The QANIMATE program allows the user to "drive" the charge as if it were a car. The **arrow** keys can be thought of as acting to increment and decrement stepper motors to turn the wheel or to push the gas pedal. While this feature of the steering may come naturally to car owners, it may be disconcerting even to them when deceleration produces negative speeds. The "acceleration" in this part of the simulation is actually the tangential component of the full acceleration. The magnitude of this tangential acceleration can be changed from a data input screen available from the pull-down menu. The other, "centripetal" component of acceleration is proportional to the rate of "turning" and is controlled by the keyboard arrow keys. A period of negative tangential acceleration will cause the velocity to pass

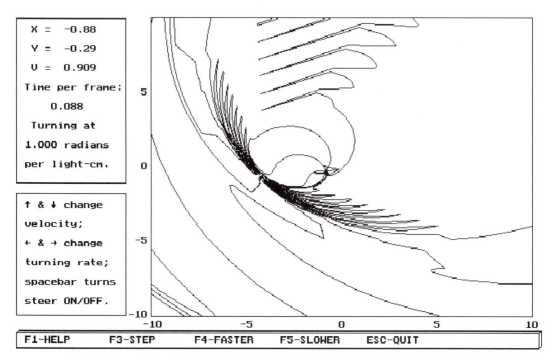

```
X =   -0.88
Y =   -0.29
U =   0.909
Time per frame:
    0.088
Turning at
1.000 radians
per light-cm.
```

```
↑ & ↓ change
velocity;
← & → change
turning rate;
spacebar turns
steer ON/OFF.
```

```
F1-HELP      F3-STEP      F4-FASTER      F5-SLOWER      ESC-QUIT
```

Figure 7.4: Sample screen from the QANIMATE program.

through zero and become negative, representing a reversal of the original direction of motion. Each press of the **up arrow** key produces an increment in the tangential velocity by an amount equal to the acceleration multiplied by the time interval between animation frames; each press of the **down arrow** key produces a velocity decrement in the same amount. The **down arrow** causes an increase in speed when the velocity is negative.

A sample screen from QANIMATE is shown in Figure 7.4. The hot keys labeled **FASTER** and **SLOWER** change the animation speed by changing the rate at which points on the field lines are generated. Pressing the **FASTER** hot key will speed up the animation, but will reduce the smoothness of the field lines. Animation speed should not be confused with the physical speed of the charge, which remains unaffected by these keys.

"Turning" is accomplished numerically by changing the direction of the motion at each time step, as is accomplished for a car through the static friction between tires and road. In the simulation, it is possible to toggle this turning on and off as if the friction between car tires and roadbed could be turned on and off at will. This latter feature allows a convenient mode of steering through simple toggling of the turning. When the charge leaves the viewport, it reappears with the same velocity entering the viewport at an opposite point. When this happens, all field lines are destroyed and must be regenerated. Since less plotting of field lines is required for the subsequent few animation frames, these frames will be "flipped" more rapidly and the animation will be faster until a full field pattern is regenerated. Therefore, it is a good idea to steer the charge so as to keep it within the viewport.

7.5.5 Displaying Wavefronts

Options in the animation part of the simulation allow alternative display modes that can be helpful in understanding the origins of the radiation field. One of these is the display of circularly expanding wavefronts centered on points on the trajectory of the charge. It is possible to display the points on these wavefronts where the electric field lines intersect them.

At high speeds ($v_{max} > 0.9$) the line segments used to draw the field lines can become too long, giving the unphysical appearance of a jumble of intersecting lines. It is useful at very high speeds either to display the wavefronts and points rather than the field lines themselves, or to slow the animation to the point that the field lines no longer appear to intersect.

7.6 Exercises

7.1 **Fields of a Point Charge in Uniform Motion**
The electric field of an unaccelerated charge can be investigated by choosing the deceleration trajectory at some viewing time $t < 0$.

   a. Examine the field pattern for several values of v_{max} to verify that the electric field for a uniformly moving point charge is a Lorentz-contracted version of the field of the charge in its rest frame. Use the **EXPLORE** menu option to measure the components of the electric field at a set of standard locations around the moving charge for various values of v_{max} to investigate the dependence of the components of $\vec{E}$ on v_{max}. Verify Eq. 7.1 by making a plot of $\tan(\Theta_{lab}) vs v_{max}$ for at least three representative values of $\tan(\Theta_{rest})$

   b. Use the **EXPLORE, GRID,** and **PLOT WHAT** menu options for some moderate value of v_{max} to investigate the other fields ($\vec{B}, \vec{A}, \Phi$, and $\vec{S}$) surrounding a point charge in uniform motion. Given that only accelerated charges produce a radiation field, explain why the Poynting vector is not zero even when there is no acceleration.

7.2 **Electric Field of an Accelerated Charge**
Use the **ANIMATION** menu option to examine the effect on the electric field of changes in both magnitude and direction of the velocity of the charge. You should first spend a few minutes becoming familiar with the use of special keys to accelerate, decelerate, and steer the charge around the screen. Notice that any deviation from uniform motion produces a curvature in the electric field lines (the signature of a radiation field).

Accelerate the charge to a constant speed of about 0.8 c, and increase the "steering" to a value that produces a moderately small circular trajectory (radius of a centimeter or two). Then use the **space bar** to toggle the steering on and off. This will allow you to guide the particle around in a trajectory at constant speed with episodes of straight-line uniform motion interspersed with circular arcs of centripetal acceleration. Leaving steering on at any point produces a purely circular trajectory. Note that every episode of acceleration produces an outward-moving pulse of radiation. While the charge is moving in a circular trajectory, increase and decrease the speed. Note that the field lines aren't smooth at high velocities far from the charge. Explain why this breakdown in the animation algorithm happens. Change the display options to show wavefronts instead of field lines and repeat the above steering sequence. Explain why the wavefronts show a spiral pattern for high-speed circular motion.

7.3   Radiation from a Linearly Accelerated Charge

The first two trajectories on the **TRAJECTORY** menu generate the fields produced when a point charge is given a uniform acceleration or deceleration for a time interval of 1 time unit. The magnitude of the acceleration in either case is v_{max} in the units employed in the program.

Choose some time after the acceleration or deceleration has stopped (say, $t = 2.0$) and use hot keys to increment or decrement v_{max} in steps of 0.05 or 0.10. The "radiation pulse" is the transition region between the field near the charge and the field farther away. Note how each line of $\vec{E}$ in the transition region connects corresponding lines of the near and far fields. Also note the way in which the electric field in the radiation pulse is affected by both the amount of acceleration and the degree of Lorentz contraction. Use the **EXPLORE** and/or **PLOT WHAT** option to investigate the dependence of the magnitude of the Poynting vector in the pulse on v_{max}. Sketch a plot of the peak value of $\vec{S}$ versus v_{max}. Is the $\vec{S}$ in the radiation pulse proportional to any simple function of v_{max}? Be sure to use a part of the pulse that is expanding in a direction perpendicular to the motion of the charge so that only the radiation field contributes to $\vec{S}$.

Next choose some appropriate value of v_{max}—say, 0.80—and use hot keys to step in increments of viewing time to watch the creation and expansion of the radiation pulse from, say $t = -1$ to some later time when the radiation pulse has left the viewing area. Use the **EXPLORE** and **PLOT WHAT** options to verify that the peak value of $\vec{S}$ has an inverse square dependence on distance from the center of the pulse.

Use the **GRID** menu option to examine the pattern of the $\vec{B}$ and $\vec{A}$ fields for both the cases of acceleration and deceleration. Give a qualitative explanation of the field pattern exhibited in each case.

7.4   Dipole Radiation

The field pattern that results from simple harmonic oscillation of a point charge changes character as the maximum speed of the charge becomes relativistic. When observed by computer simulation, this transition turns out to be a surprisingly rapid function of charge speed. In the simple harmonic motion trajectory, the amplitude of the motion is fixed at unity, so the frequency increases in proportion to v_{max}, and the peak acceleration increases in proportion to the square of v_{max}.

Examine the changes in the electric field pattern from a charge in linear simple-harmonic motion when the value of v_{max} is varied from, say, 0 to

0.95 in increments of 0.05. Note the change in character of the field lines from gentle curvature to sharp bends as the velocity and acceleration increase. Use the **EXPLORE, GRID,** and **PLOT WHAT** menu options to investigate the changes in behavior of the radiation field as v_{max} increases to relativistic values (approaches unity). Use the **PLOT WHAT** feature to measure the wavelength of the radiation for smaller values of v_{max} and verify that the wavelength is inversely proportional to frequency. (Note that v_{max} is the angular frequency in these units.)

In the non-relativistic limit ($v \ll c$), the intensity of the radiation from a charge in simple harmonic motion should be zero along the axis of motion and a maximum in the direction perpendicular to the motion; this is the double-lobed angular distribution characteristic of dipole radiation. Also, the dipole radiation should be approximately sinusoidal with the same frequency as the simple harmonic motion. Verify that this is true using the built-in trajectory.

Look at plots of $\vec{S}$ along cuts that contain several wavelengths, and use the **EXPLORE** option to measure peak values (you may need to enlarge the plotting scale for low values of v_{max}). Is the dipole description appropriate when the particle speed becomes relativistic? Sketch the angular distribution of radiation from a linear oscillator in the relativistic limit, again by examining peak values of $\vec{S}$ at the same distance in different directions. Comment on the differences with the non-relativistic case.

7.5   Cyclotron and Synchrotron Radiation

The uniform circular motion trajectory is an example of continuous harmonic charge acceleration that is useful for examining the transition from a monochromatic radiation field (cyclotron) to the broad-band radiation (synchrotron) that occurs as the motion of the charge becomes relativistic. It is also relatively easy to create uniform circular motion in the animated part of this simulation by steering to maintain a fixed "turning rate."

Use the still-frame part of the simulation and hot keys to step the circular motion trajectory incrementally from $v_{max} = 0.1$ to $v_{max} = 0.95$ to observe this change in character of the radiation with increasing charge speed. Use the **GRID, EXPLORE,** and **PLOT WHAT** options to investigate the relationship between $\vec{B}$ and $\vec{E}$ at both low and high charge speeds.

Since the radiation field spirals outward at speed c, a plot of any of the fields along a short radial cut will be roughly equivalent to what a stationary observer would record for that field as a function of time. Use this technique and the **PLOT WHAT** menu option to investigate the time dependence of the different fields as the charge speed changes incrementally from $v_{max} = 0.1$ to $v_{max} = 0.95$.

The square of the Fourier transform of the time-dependent field is proportional to its frequency spectrum. Verify that the synchrotron radiation is pulsed and broad band, while the cyclotron field is continuous and monochromatic. This can be done using several of the built trajectories by examining a plot of the shape of $\vec{S}$ along a radial cut away from the source of the radiation. Is the shape of the synchrotron pulse different for the five different fields that can be plotted using the **PLOT WHAT** option?

7.6   Relativistic Beaming and the "Headlamp Effect"

The **SINUOUS** option on the **TRAJECTORY** menu combines linear motion in one direction with simple harmonic motion in a perpendicular direction. Motion of this type by a relativistic beam of electrons in a "wiggler accelerator" has been proposed as a directional source of synchrotron radiation.

Investigate the dependence of the radiation field from a charge in sinuous motion on v_{max} as the maximum speed changes from $v_{max} = 0.1$ to $v_{max} = 0.95$. Use the **PLOT WHAT** and **EXPLORE** menu options to compare the Poynting vector along the forward and backward directions for several values of v_{max}. Sketch a plot of the ratio of the peak intensity of the radiation in the forward direction to that in the backward direction versus v_{max}. Formulate a qualitative explanation of this forward beaming of radiation.

7.7   The Doppler Effect

Use the **PLOT WHAT** option to measure both the wavelength and the period of the radiation in the forward and backward directions for the the **SINUOUS** trajectory at some relatively small value of v_{max} (say, 0.20). Verify that the ratios of the wavelengths and periods are about what would be predicted by non-relativistic Doppler effect formulas. (It may be necessary to adjust the scale on the field viewport to get an accurate measure of the wavelengths.) Given that $c = 1$, what is the ratio of v_{max} to the average of the two corresponding frequencies? Repeat the measurement for a large value of v_{max} (say, 0.90). Is the ratio of v_{max} to the average of the frequencies the same as in the non-relativistic case? Are your results consistent with relativistic Doppler effect formulae? How is the forward component of the charge velocity related to the value of v_{max}?

7.8   The Vector Potential

The vector potential $\vec{A}$ produced by a moving charge is in the direction of the velocity of the charge at the retarded time. Use the **GRID** option to display the vector potential for the cases of LINEAR ACCELERATION and LINEAR DECELERATION. Note the difference between the patterns inside and outside the radiation pulse. Repeat for the 90-degree turn trajectory.

The magnetic field at each point in space is the *curl* of the vector potential. Display a grid of $\vec{B}$ to verify that the magnetic field in each of the above cases is in the direction of the circulation of $\vec{A}$, i.e., that a right-hand screw would move in the direction of $\vec{B}$ if turned in the direction $\vec{A}$ circulates.

Acknowledgments

Mr. Gregory Meyers did much of the original programming for the interactive animation portion of this simulation, and contributed several insightful ideas as well, while he was a high-school student enrolled in the summer Research Apprenticeships in Science Program at Bowling Green State University. Both the manuscript for this chapter and the accompanying simulations were greatly improved by suggestions from several students, colleagues, and others who reviewed and tested them in their classes.

Bibliography

1. Purcell, E. M. *Berkeley Physics Course*. Vol. 2, Chapt. 5: Electricity and magnetism. New York: McGraw Hill, 1965.

2. Lorrain, P., Corson, D. P., and Lorrain, F. *Electromagnetic Fields and Waves*. Chapt. 8, San Francisco: Freeman Press, 1988.

3. Hamilton, J. C. and Schwartz, J. L. Electric fields of moving charges: A series of four film loops. American Journal of Physics *39*:1540, 1971.

4. Good, R. H. Simulation Programs. Computers in Physics SEP/OCT:76, 1988.

5. Cabrerra, B. "Radiation," Physics Simulations, Vol. 2. Intellimation Library for the MacIntosh.

6. Tsien, R. Y. Pictures of dynamic electric fields. American Journal of Physics, underbar 40:48, 1972.

7. Jackson, J. D. *Classical Electrodynamics*. Chapt. 14: Radiation by Moving Charges. New York: John Wiley and Sons, 1962.

8. Marion, J. B. *Classical Electromagnetic Radiation*. Chapt. 7: The Lienard-Wiechert potentials and radiation. New York: Academic Press, 1965.

8

Electromagnetic Waves

Ronald Stoner

"Gentlemen," said Candide, "as to the plenum, it is uncontestable that there is no vacuum in nature, and the *material subtilis* is well imagined." "You are impertinent, friend," replied the philosophers... "Have you any notion of the theory of light?"

—Voltaire

8.1 Introduction

The presence throughout space of the universal microwave background demonstrates that electromagnetic radiation can carry energy and information over truly astronomical reaches of space and time. The persistence of electromagnetic waves far from their source is possible because the electric and magnetic fields reinforce each other in their respective oscillations: Time variation in the electric field is the "displacement current" source of the magnetic field, and time variation of the magnetic field produces the electric field by Faraday induction.

The fields that constitute electromagnetic radiation are difficult to visualize or to portray in the static, two-dimensional format of textbook line drawing, because the precise ways the electric and magnetic fields interact in radiation is inherently three-dimensional and dynamic. Computer simulation and animation provide a way to visualize the dynamic, three-dimensional character of their interaction. This CUPS simulation uses perspective projection and animation to allow the user to visualize various polarization states of coupled harmonic waves of electric and magnetic fields, both in free space and in matter. The simulation also can be used to illustrate the phenomena of reflection and transmission at a boundary, to observe differences between traveling and standing waves, and to investigate the effects of optical elements such as polarizers and quarter-wave plates.

8.2   Electromagnetic Waves in Vacuum

Maxwell's equations govern the behavior of the electric field $\vec{E}$ and the magnetic field $\vec{B}$ in the classical regime of electromagnetism. In vacuum, these equations take the following simple form:

$$\vec{\nabla} \cdot \vec{E} = 0 , \tag{8.1}$$

$$\vec{\nabla} \times \vec{E} = \frac{-1}{c} \frac{\partial \vec{B}}{\partial t} , \tag{8.2}$$

$$\vec{\nabla} \cdot \vec{B} = 0 , \tag{8.3}$$

$$\vec{\nabla} \times \vec{B} = \frac{1}{c} \frac{\partial \vec{E}}{\partial t} . \tag{8.4}$$

The above form of Maxwell's equations is appropriate when cgs (Gaussian) units are used for the fields,[1] and are convenient for treating electromagnetic wave phenomena; the form of Maxwell's equations appropriate for the case of rationalized mks units can be found in standard textbooks.[2] In the cgs system, $\vec{E}$ and $\vec{B}$ have the same units (Gauss), which has particular advantages for the analysis of electromagnetic radiation.

Among the particular solutions of the free-space Maxwell's equations are those in which both the electric field, $\vec{E}(\vec{r}, t)$, and the magnetic field, $\vec{B}(\vec{r}, t)$, have the mathematical forms of polarized, traveling plane waves, namely,

$$\vec{E}(\vec{r}, t) = \vec{E}_o e^{i(\omega t - \vec{k} \cdot \vec{r})} \tag{8.5}$$

$$\vec{B}(\vec{r}, t) = \vec{B}_o e^{i(\omega t - \vec{k} \cdot \vec{r})} . \tag{8.6}$$

The symbols $\vec{E}_o$ and $\vec{B}_o$ in these plane-wave solutions are the amplitudes of a plane harmonic wave, which are vectors with complex components. The use of complex numbers and exponential functions allows the mathematical description of the harmonic waves to be done more compactly. It is implicitly understood that the actual fields are the real parts of the complex functions on the right-hand sides of the Eqs. 8.5 and 8.6. The $\vec{k}$ in the above plane-wave formulas is the wave vector ($|\vec{k}| = \frac{2\pi}{\lambda}$), and $\omega = 2\pi f$ is the angular frequency of the wave.

Since Maxwell's equations in matter-free space admit plane-wave solutions, they can be thought of as the wave equations for electromagnetic waves in space. Since the equations connect partial space derivatives of one field with partial time derivatives of the other, the plane waves for $\vec{E}(\vec{r}, t)$ and $\vec{B}(\vec{r}, t)$ are not independent. The parameters in the plane-wave description of $\vec{E}(\vec{r}, t)$ completely determine the parameters in the $\vec{B}(\vec{r}, t)$ plane wave. The connection can be established by substitution of the traveling plane-wave forms into Maxwell's equations, which reveals that they represent possible solutions only under the following conditions:

$$\vec{k} \cdot \vec{E}_o = 0 \tag{8.7}$$

$$\vec{k} \times \vec{E}_o = \frac{\omega}{c} \vec{B}_o \tag{8.8}$$

$$\vec{k} \cdot \vec{B}_o = 0 \tag{8.9}$$

$$\vec{k} \times \vec{B}_o = -\frac{\omega}{c} \vec{E}_o \tag{8.10}$$

These mathematical conditions imply that the three vectors $\vec{k}$, $\vec{E}_o$, and $\vec{B}_o$ are mutually perpendicular, and that the magnitudes of $\vec{E}_o$ and $\vec{B}_o$ are equal. They further imply that $\omega = ck$, which means that electromagnetic waves in space are dispersionless, and that the phase velocity of the waves is $\lambda f = c$, the speed of light.

It is always possible to choose a frame of reference in which $\vec{k}$ is in the $+z$-direction, in which case $\vec{E}_o$ and $\vec{B}_o$ have only x- and y-components. The restrictions that the magnitudes of these two vector amplitudes are equal and that the two vectors are mutually perpendicular still leave considerable latitude, since each has two complex components. It is the relative amplitudes and phases of the two Cartesian components of $\vec{E}_o$ that determine the polarization of the wave. The conventions and symbols used below to represent this polarization are those used in the standard advanced text by Born and Wolf.[3] It is always possible to choose the origin and the zero of time in such a way that the real parts of the two components of $\vec{E}$ can be written

$$E_x = a_1 \cos(\omega t - \vec{k} \cdot \vec{r}) \tag{8.11}$$

and

$$E_y = a_2 \cos(\omega t - \vec{k} \cdot \vec{r} + \delta), \tag{8.12}$$

where a_1 and a_2 are real amplitudes, and where δ is the difference in phase between the x- and y-components of the electric field. In this case, the complex values of the two components of $\vec{E}_o$ and $\vec{B}_o$ are

$$E_{ox} = a_1, \tag{8.13}$$

$$E_{oy} = a_2 e^{i\delta}, \tag{8.14}$$

$$B_{ox} = a_2 e^{i\delta}, \tag{8.15}$$

and

$$B_{oy} = -a_1. \tag{8.16}$$

The parameter δ represents the radian measure of the difference in phase of oscillation of the two components of $\vec{E}$ (or of $\vec{B}$). If $\delta = 0$, then the two components oscillate in phase and the wave is plane polarized. If $\delta = \frac{\pi}{2}$, then the y-component of $\vec{E}$ comes to a maximum one-fourth of a period before the x-component, and the radiation is said to be right-circularly polarized (i.e., negative helicity). If $\delta = -\frac{\pi}{2}$, then the x-component of $\vec{E}$ comes to a maximum one-fourth of a period before the y-component, and the radiation is said to be left-circularly polarized (positive helicity). If $\delta = \pi$ (or $-\pi$), the wave is again plane polarized, but in a direction perpendicular to the $\delta = 0$ case. Intermediate values of δ correspond to various degrees of "elliptical" polarization.

8.3   Stokes' Parameters

Maxwell's equations and the particular choices for the origins of space and time leave three degrees of freedom for describing a monochromatic electromagnetic plane wave. These three degrees of freedom correspond to the freedom to choose values of the three parameters a_1, a_2, and δ. In the case of fully polarized, coherent electromagnetic radiation, these three parameters can be related to measurable quantities that are traditional for characterizing the polarization state of electromagnetic radiation. These quantities are the Stokes' parameters,[3,4] which can be related to a_1, a_2, and δ as follows:

$$s_0 = a_1^2 + a_2^2,\tag{8.17}$$

$$s_1 = a_1^2 - a_2^2,\tag{8.18}$$

$$s_2 = 2a_1a_2\cos(\delta),\tag{8.19}$$

$$s_3 = 2a_1a_2\sin(\delta).\tag{8.20}$$

There are traditionally four independent Stokes' parameters measured in the laboratory to characterize the polarization state of electromagnetic radiation, but only three of the four parameters defined above are truly independent, since they are derived from only the three free parameters a_1, a_2, and δ. The fourth degree of freedom corresponds to the "degree of polarization" of a "partially polarized" beam of radiation. In the case of partially polarized radiation, the quantities a_1, a_2 and δ are not constant, but functions of time that change very slowly over many, many periods of wave oscillation. But a_1, a_2, and δ may be quite variable over the much longer time scales required to measure the Stokes' parameters in the laboratory. This means that Stokes' parameters measured in the laboratory are actually time averages, at least in the partially polarized case.

For similar reasons, the mathematical description and computer-assisted visualization of partially polarized radiation is beyond the scope of this simulation. This is because partially polarized radiation cannot be truly monochromatic and coherent. While it may be possible to think of a plane wave whose polarization state changes slowly with time, the Fourier spectrum of such a wave would necessarily have a finite frequency width. This frequency width corresponds to the rate at which the polarization state changes. In this sense, partially polarized radiation is only "quasi-monochromatic."

In the laboratory, quasi-monochromatic radiation is produced by noncoherent sources and has a very narrow frequency spectrum. The narrowness of the frequency distribution implies that the radiation is coherent over a very large number of periods of oscillation. It is effectively impossible to simulate such quasi-monochromatic radiation by the computer animation technique employed here, at least with any degree of realism, since to do so would require generating an extremely large number of periods or wavelengths simultaneously. For that reason, the present simulation is confined to the treatment of fully polarized (i.e., exactly monochromatic) radiation, meaning that the quantities a_1, a_2, and δ are constant in time. Therefore the Stokes' parameters are not independent, but subject to the condition that

$$s_0^2 = s_1^2 + s_2^2 + s_3^2.\tag{8.21}$$

Equations 8.17–8.20 are compatible with Eq. 8.21, which is the constraint that reduces the degrees of freedom from four to three for "fully polarized" or "purely monochromatic" radiation.

All four Stokes' parameters have the dimension of intensity, i.e., radiative power per unit area. The overall intensity of a plane electromagnetic wave is given by the magnitude of the Poynting vector $\vec{S} = c\vec{E} \times \vec{B}$, which, averaged over one period of oscillation, is $\frac{c}{2}(a_1^2 + a_2^2)$. It follows that the Stokes' parameter s_0 is a measure of the overall intensity of the electromagnetic wave. The other three Stokes' parameters have the following meanings:

- s_1 represents the difference between the intensity transmitted by a linear polarizer with polarizing axis in a given direction (say, 0°) and the intensity transmitted when the polarizer is rotated to a perpendicular direction (90°);

- s_2 is similar to s_1 except that the mutually perpendicular directions of the polarizer axis are 45 degrees and 135 degrees; and

- s_3 represents the difference between the intensity of the radiation after passing through a right-circular polarizer and its intensity after passing through a left-circular polarizer.

8.4   Electromagnetic Waves in Matter

When matter is subjected to electric and magnetic fields, electric and magnetic dipoles are induced in the individual atoms. The electric and magnetic polarization of the atoms represents microscopic electric charges and electric currents that themselves are sources of electric and magnetic fields. Over distance scales much larger than atomic sizes, these induced dipoles appear as continuous electric polarization and magnetization fields in the material, and the microscopic charges and currents appear as macroscopic distributions of "bound" electric charge and current. Further, an electric field applied to a material with mobile microscopic charge carriers, such as the free electrons in metals, will generate "free" electric currents that can be described by a electric current density $\vec{J}(\vec{r}, t)$.

The effect of the applied fields in most materials is "linear," which means that the polarization fields and free currents are proportional to the fields that induce them. Since all these induced charges and currents are both produced by the fields and are themselves sources of the fields, they are coupled with the electromagnetic wave. They oscillate coherently with the electric and magnetic fields, they carry part of the energy and momentum, and so in a real sense are part of the electromagnetic wave in the material.

It is often mathematically convenient in the presence of matter to introduce two new macroscopic fields, $\vec{D}(\vec{r}, t)$ and $\vec{H}(\vec{r}, t)$, and to use them to rewrite Maxwell's equations in such a way that bound charges and bound currents do not appear as explicit sources of $\vec{E}(\vec{r}, t)$ and $\vec{B}(\vec{r}, t)$. This version of Maxwell's equations, in the absence of free electric charge, is[1,2]

$$\vec{\nabla} \cdot \vec{D} = 0, \tag{8.22}$$

$$\vec{\nabla} \times \vec{E} = \frac{-1}{c} \frac{\partial \vec{B}}{\partial t}, \tag{8.23}$$

$$\vec{\nabla} \cdot \vec{B} = 0, \tag{8.24}$$

$$\vec{\nabla} \times \vec{H} = \frac{1}{c} \frac{\partial \vec{D}}{\partial t} + \frac{4\pi}{c} \vec{J}, \tag{8.25}$$

where $\vec{J}$ is the density of free current.

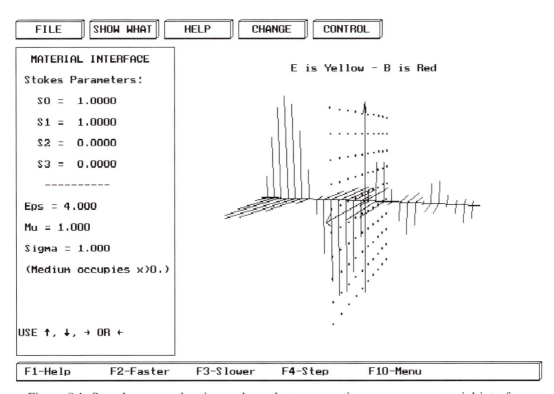

Figure 8.1: Sample screen showing a plane electromagnetic wave near a material interface.

Figure 8.1 shows a sample screen from section of this simulation that depicts the behavior of electromagnetic waves incident from free space onto an isotropic, homogeneous, linear material.

In linear isotropic materials, the atomic polarizations are both parallel to and proportional to the applied $\vec{E}$ and $\vec{B}$ fields, and Ohm's law describes the free electric current produced by an applied electric field. Maxwell's equations can be further simplified in this case by writing

$$\vec{D} = \epsilon\vec{E}, \tag{8.26}$$

$$\vec{B} = \mu\vec{H}, \tag{8.27}$$

and

$$\vec{J} = \sigma\vec{E}, \tag{8.28}$$

where ϵ, μ, and σ are real constants that characterize the material. They have the following meanings:

- ϵ is the relative electric permittivity,

- μ is the relative magnetic permeability, and

- σ is the electrical conductivity.

In cgs units, ϵ and μ are dimensionless pure numbers, identical to the relative permittivity and permeability in rationalized mks units. But $\vec{E}$ and $\vec{J}$ are of different dimension, so Ohm's law (Eq. 8.28) defines the conductivity σ to have the dimension of current density per unit field. This is inconvenient for computation, and it makes the numerical value of σ depend on the system of units chosen. For those reasons, the simulation employs and asks the user to supply a dimensionless σ that is proportional to the real conductivity of the material. More specifically, in cgs units, σ has the dimension of frequency; however, ω is fixed in the simulation. The code defines $\omega = 1$ implicitly for computational efficiency, while the apparent "real time" frequency of oscillation is fixed by computer clock speed and the user-supplied time increment between animation frames. The implicit unit adopted for σ is $\frac{\omega}{4\pi} = f/2$. Therefore, to simulate a particular material and frequency, the user should substitute the dimensionless ratio 2(conductivity/frequency) for the σ in the data input screen.

If the material is homogeneous, ϵ, μ, and σ are uniform in space and have no space derivatives inside the material. In this case, Maxwell's equations can again be written in terms of $\vec{E}$ and $\vec{B}$ as follows:

$$\vec{\nabla} \cdot \vec{E} = 0, \tag{8.29}$$

$$\vec{\nabla} \times \vec{E} = \frac{-1}{c} \frac{\partial \vec{B}}{\partial t}, \tag{8.30}$$

$$\vec{\nabla} \cdot \vec{B} = 0, \tag{8.31}$$

$$\vec{\nabla} \times \vec{B} = \frac{\epsilon\mu}{c} \frac{\partial \vec{E}}{\partial t} + \frac{4\pi\mu\sigma}{c} \vec{E}. \tag{8.32}$$

Implicit in these equations is the assumption that ϵ, μ, and σ are real constants, which is a valid assumption only if the current density and polarization fields respond immediately to changes in the applied fields. In other words, the equations are valid if the time scales for variations of the fields are much longer than both the natural periods of atomic oscillations and the time intervals between scatterings of the free charge carriers in the material.

Equations 8.29–8.32 are approximations used in this simulation for the partial differential equations of electromagnetic waves in matter. Substitution of the plane-wave forms of Eqs. 8.5 and 8.6 into them produces the following condition on the wave vector k and frequency ω of the plane wave:

$$k^2 = \frac{\omega^2}{c^2} \left(\mu\epsilon + i\frac{4\pi\mu\sigma}{\omega} \right), \tag{8.33}$$

where ω is real by virtue of the implicit condition that the fields at a given point in space are simple harmonic functions of time. The imaginary part of the right-hand side of this equation represents the effect of the current density term in Maxwell's equations. The oscillations of the current density are $\frac{1}{4}$ period out of phase with the other fields, which provides a mechanism for the dissipation of the energy carried by the wave. The presence of this term means that k will have both real and imaginary parts:

$$k = \beta + i\alpha, \tag{8.34}$$

where both β and α depend on the material properties ϵ, μ, and σ, and on the frequency ω. The imaginary part, α, leads to the exponential attenuation of the wave along the direction of propagation, and the real part, β, is the phase gradient in the material, i.e., $\beta = \frac{2\pi}{\lambda}$.

8.4.1   Reflection and Transmission at a Boundary

Straight lines perpendicular to the axis of propagation, the z-axis, are used throughout this simulation to represent the fields $\vec{E}$ and $\vec{B}$ evaluated at points along that axis. Because the x- and y-dimensions are used this way to represent the magnitudes and directions of the fields, it isn't feasible to use the x- and y-dimensions to convey additional information, such as changes in direction of a plane wave upon reflection or refraction at an interface. On the other hand, reflection and transmission of a wave normally incident on a plane boundary occurs along the axis of propagation, so it can be represented.

When a plane electromagnetic wave is incident on the interface between free space and matter, Maxwell's equations must be satisfied at the interface. This leads to boundary conditions that relate $\vec{E}$ and $\vec{B}$ on opposite sides of the interface. The boundary conditions require, in general, that the incident wave be partially reflected and partially transmitted. For example, the requirement that $\vec{\nabla} \times \vec{E} = \frac{-1}{c}\frac{\partial \vec{B}}{\partial t}$ at the interface, together with the physical requirement that $\frac{\partial \vec{B}}{\partial t}$ be finite, requires for normal incidence that neither $\vec{E}$ nor $\frac{\partial \vec{E}}{\partial t}$ change discontinuously across the boundary. The abrupt change in magnetization at the interface means that there is no similar requirement that $\vec{B}$ be continuous.

Consider the special case of an incident wave with $\vec{k} = k\hat{z}$. This wave will be transmitted across an interface lying in the x-y plane into the medium with the same frequency ω, but with a generally different wave vector $\vec{k}_t = k_t\hat{z}$. In order to match the above boundary conditions on $\vec{E}$, the incident wave must also be partially reflected from the boundary, leading to a reflected plane wave with $\vec{k}_r = -k\hat{z}$. The boundary conditions on the electric field then connect the superposition of incident and reflected wave on one side of the boundary with the transmitted wave on the other.

The code for the simulation of reflection and transmission at a material interface proceeds in the following way: The Stokes' parameters of the incident plane wave determine the parameters a_1, a_2, and δ for it through the inverse of Eqs. 8.17–8.20. The parameters ϵ, μ, and σ are then used to find the real and imaginary parts of k_t. Next, the boundary conditions on $\vec{E}$ are used to determine the amplitudes and phases of both the x- and y-components of $\vec{E}$ and $\vec{B}$ of the reflected and transmitted waves. This finally allows $\vec{E}$ and $\vec{B}$ to be computed at all times t and positions z along the z-axis.

8.5   Polarizers and Quarter-Wave Plates

Polarizers are devices that resolve incident electromagnetic radiation into components that are parallel and perpendicular to the optical axis of the polarizer, then transmit only the parallel component. Quarter-wave plates are devices that resolve incident radiation into components parallel and perpendicular to an optical axis, then introduce a relative phase shift of 90 degrees ($\frac{\pi}{2}$ radians) before combining and transmitting them again.[4] This may be accomplished in different ways in

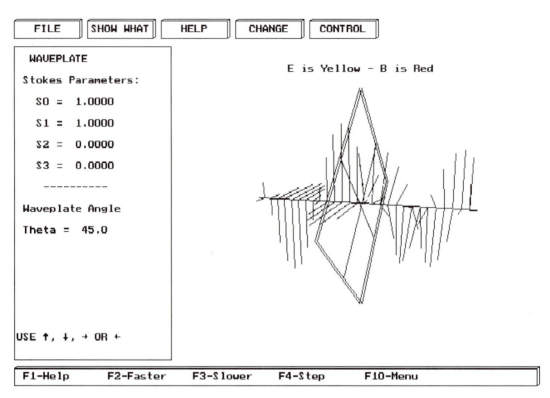

Figure 8.2: Sample screen showing conversion of a plane polarized wave into a circularly polarized wave by a quarter-wave plate.

different frequency bands, but in the case of visible light, polarizers and quarter-wave plates are many wavelengths thick. In the present simulation these devices are represented by drawings of very thin square plates.

Depending on the orientation of its optical axis, a perfect quarter-wave plate can convert a plane-polarized plane wave to a circularly polarized wave without change in intensity, and vice versa. Figure 8.2 is a sample screen that depicts this process.

A polarizer can also convert an incident elliptically or circularly polarized wave to a transmitted plane-polarized wave, but this generally involves a decrease in intensity since only one of the two components of the incident wave is transmitted.

The code for this simulation computes the effect of a polarizer in the following way: As in the other cases simulated, the polarization of the incident wave is characterized by Stokes' parameters, from which a_1, a_2, and δ can be computed for the incident wave. These parameters are used to create a two-component vector representing the complex amplitude $\vec{E}_o = (a_1, a_2 e^{i\delta})$. This two-component vector is successively multiplied by three 2×2 matrices representing the following operations: (1) rotation to bring the plane of polarization parallel to the optic axis of the polarizer; (2) removal of the component of $\vec{E}_o$ perpendicular to the optic axis; and (3) rotation of the plane of polarization back to the original direction.

The result is a two-component vector for the complex amplitude of the transmitted wave, $\vec{E}_{ot}$, which is then converted to parameters a_{1t}, a_{2t}, and δ_t for the transmitted wave.

A very similar procedure is used to represent the quarter-wave plate, the only difference being that the second matrix multiplication, instead of removing the *y*-component, introduces a quarter-wave phase shift by multiplying it by $e^{i\frac{\pi}{2}} = i$. The effects of various combinations of polarizers and wave plates can be treated in much the same way through successive matrix multiplication.

8.6   *Using the Simulation*

Each frame of the simulation shows a perspective projection of coordinate axes and color-coded vectors representing $\vec{E}$ and $\vec{B}$ at uniformly spaced intervals along the *z*-axis. Animation is achieved by frame flipping. While one frame is displayed, another hidden one is being drawn that corresponds to a slightly later time. Once the drawing of the hidden frame is complete, it is displayed, the previous frame is hidden, erased, and redrawn to show the fields at a still later time.

The perspective projection is drawn with the origin at screen center and from the point of view of an observer a fixed distance from the origin. Pressing the arrow keys moves the observation point in angular increments of 5 degrees in altitude and azimuth relative to the *x-z* plane. This allows the user to examine the wave from many different points of view so as to see its three-dimensional structure.

The ability to change the user's viewpoint can be used effectively to examine a single component of the fields: Viewing the wave from a point on the *x*-axis will yield a projection of the *x*-components of the fields on the screen, and viewing from a point on the *y*-axis will produce a projection of the *y*-components. Similarly, the nature of the polarization (plane, circular, or elliptical) may be most obvious by viewing it from a point on either the positive or negative *z*-axis.

Some of the simulations have a hot key labeled **Rotate**, which allows the user to change the orientation of a polarizer or waveplate in 5 degree increments "on the fly." This has the pedagogic advantage of allowing "real time" simulation of optics exercises. Each such shift in orientation of a polarizer will break the coherence of the simulated wave, and may result in an instantaneous large jump in phase. This is not unphysical since it would be impossible in practice to rotate a polarizer by any significant amount during a fraction of one oscillation of an optical wave. Similar phase jumps may result when using hot keys labeled **Faster** and **Slower**.

8.7   *Exercises*

8.1   **The Stokes' Parameters**

The first Stokes' parameter is fixed at $s_0 = 1$ in this simulation, and the other three are automatically normalized to satisfy $s_0^2 = (s_1^2 + s_2^2 + s_3^2)$. Use the simulation to find and record the values of sets of the parameters (s_1, s_2, and s_3) that produce each of the following types of polarization:

a. Plane polarization in the *x*-direction

b. Plane polarization in the *y*-direction

    c. Plane polarization at a 45 degree angle with the x-direction;

    d. Right circular polarization;

    e. Left circular polarization; and

    f. Elliptical polarization with semimajor axis double that of the semiminor axis.

8.2   Are $\vec{E}$ and $\vec{B}$ in Phase?

Examine the phase relationship between $\vec{E}$ and $\vec{B}$ for a plane polarized wave ($s_3 = 0$). Note that they oscillate in phase. Show that this is required by Maxwell's equations (Eqs. 8.1–18.4) because space derivatives of $\vec{E}$ are proportional to time derivatives of $\vec{B}$ (and vice versa). Next examine a circularly polarized wave ($s_1 = s_2 = 0$, $s_3 = 1$) from a viewpoint along either the x-axis or the y-axis. Explain why $\vec{E}$ and $\vec{B}$ appear to be 90 degrees out of phase from this perspective, and why this is consistent with Maxwell's equations.

8.3   The Speed of Electromagnetic Waves in Matter

The speed (phase velocity) of a harmonic wave is the wavelength times the frequency: $v_{phase} = f\lambda = \frac{\omega}{k}$. Use the simulation to measure the wavelength of a plane wave in insulating matter (i.e., $\sigma = 0$) for several values of μ and ϵ. This can be done by choosing a viewpoint far from the origin and in the y-z plane, then measuring distance with a ruler on the computer screen. Given that the frequency in all cases is the same, and given that the speed of radiation in free space is c, determine the speed of the wave in matter in each case. Plot $log(v_{phase})$ vs. $log(\mu)$ for $\epsilon = 1$ and $log(v_{phase})$ vs. $log(\epsilon)$ for $\mu = 1$. From your plot, infer a formula for v_{phase} as a function of μ and ϵ. Given that neither energy nor information can travel faster than c, do you believe it is possible for the speed of an electromagnetic wave in matter to be larger than c?

8.4   Boundary Conditions at an Interface

Satisfy yourself by watching the simulation that $\vec{E}$ varies continuously across the interface between free space and matter for all values of μ, ϵ, and σ. Try values of μ and ϵ in the range 0.5 to 20.0, and values of $\sigma > 0$. Also, in the case $\mu = 1$, $\epsilon = 1$, and $\sigma = 0$ there is no discontinuity, so $\vec{B}$ is also continuous across the boundary in that special case. Can you find any other sets of μ, ϵ, and σ that allow $\vec{B}$ to vary continuously across the interface? Given that $\vec{B} = \mu\vec{H}$, can you determine conditions under which $\vec{H}$ is continuous across the interface?

8.5   Reflection, Transmission, and Standing Waves

Matter-free space in this simulation corresponds to $\mu = 1$, $\epsilon = 1$, and $\sigma = 0$; if the material occupying the region $z > 0$ has different values of these parameters, a plane wave incident on the interface will be partially reflected and partially transmitted into the medium beyond the interface. In the latter case, the superposition of incident and reflected waves in the region $z < 0$ produces a partially standing wave. The amplitude of the reflected wave can be inferred by noting the sizes of the maxima and minima in the standing wave envelope for a plane-polarized incident wave. These maxima and minima are, respectively, the sum and difference of the amplitudes of the incident and reflected waves.

a. Make estimates of the amplitude of the reflected wave for enough values of μ, ϵ, and σ to sketch a plot of the reflection coefficient (defined here to be the ratio of the reflected amplitude to the incident amplitude) as a function of each one of these parameters, in each case leaving the other two at their free-space values.

b. When seismic waves are reflected from interfaces between two layers of rock in geologic studies, the reflection coefficient depends only on the ratio of the speeds of sound on either side of the interface. Make measurements of the reflection coefficient and wavelength in the medium for various values of μ and ϵ to see whether the reflection coefficient for electromagnetic waves at the interface depends only on the speed of the wave inside the medium. If so, make a plot of reflection coefficient vs. speed of the wave in the medium.

c. For $\sigma = 0$, sketch a plot of the transmission coefficient (ratio of the transmitted amplitude to the incident amplitude) as a function of μ and of ϵ.

d. For very large values of μ, ϵ, and σ the incident wave is totally reflected. Compare the wave pattern produced in the region $z < 0$ when the incident wave is plane polarized with the pattern produced when the incident wave is circularly polarized. Make a sketch of the energy density $(E^2 + B^2)/2$ as a function of z for each of these cases.

8.6 Skin Depth

For $\sigma > 0$, the transmitted wave decays exponentially with distance into the region $z > 0$. That is, the amplitude of the oscillations in $\vec{E}$ varies as $e^{\frac{-z}{\delta}}$, where δ is called the "skin depth" or "penetration depth." By varying the values of μ, ϵ, and σ determine the approximate functional dependence of δ on these three quantities. Sketch rough plots of δ vs. each quantity.

8.7 Reflection by Real Materials

Look up the permittivity, permeability, and conductivity of some real materials like glass, semiconductors and metals. Convert the conductivity to cgs units ($\sec^{-1}$). Also look up the reflectivity of the same materials for radiation at different frequencies. Simulate the reflection and transmission of radiation from these materials to see if the model used to create the simulation is roughly accurate. Try to explain any differences in terms of the validity of the approximations made in section 8.4.

8.8 How to Measure the Stokes' Parameters

a. Find an orientation of a quarter-wave plate that will convert plane-polarized radiation to right circularly polarized radiation.

b. Find an orientation of a quarter-wave plate that will convert plane-polarized radiation to left circularly polarized radiation.

c. Verify that s_1 and s_2 can be measured using a single polarizer arranged at angles 0, 45, 90, and 135 degrees. Remember that intensity is proportional to the square of the amplitude.

d. Construct a scheme for measuring s_3 and use the simulation to verify that it works.

References

1. Lorrain, P., Corson, D. P., and Lorrain, F. *Electromagnetic Fields and Waves.* Chaps. 27 and 28. San Francisco: Freeman Press, 1988.

2. Jackson, J. D. *Classical Electrodynamics.* Chapter 7: Plane electromagnetic waves and wave propagation. New York: John Wiley and Sons, 1975.

3. Born, M. and Wolf, E. *Principles of Optics.* 4th edition. Chaps. 1 and 10. Elmsford, NY: Pergamon Press, 1970.

4. Guenther, R. D. *Modern Optics.* Chaps. 2 and 13. New York: John Wiley and Sons, 1990.

Appendix A

Walk-Throughs for All Programs

A.1 Walk Through: FIELDS Program

The initial screen comes up with a default choice for a scalar field: $F = xy + zr$. The four screens show F, ∇F, $\nabla^2 F$, and a "mouse probe window," which shows the values of all quantities at the current location of the mouse.

- **Select Read Scalar Field under the Field menu.**

- **Select F(r,theta,z) (spherical coordinates), and enter the expression:** $sin(4 * theta) * exp(-r^2)$. (Note that, in contrast to the default field, the Laplacian of this field is no longer zero everywhere.)

- **Examine the 3-D representation of any plot, by clicking on the small square in the upper left corner of any plot.**

- **Select Choose Graphs under the Graphs menu.**

- **Change graph 3 to the Path Integral window.**

- **With the mouse in the upper left window click the left mouse button to define corners of a path and evaluate the line integral for the selected path.** You might try integrals along the two coordinate axes.

- **Switch to a Closed path, and see if you are dealing with a conservative field by evaluating the integral for several paths.** Of course, since it is a scalar field, it will certainly be conservative.

- **Select Vector Field under the Field menu, and enter the vector field** $A_x = y$, $A_y = -x$, **and** $A_z = 0$. (Note that the curl of this field is everywhere $-2\hat{z}$, and the divergence is zero, as expected.)

- **Integrate the vector field along the line extending from** $(x, y) = (-.75, .75)$ **to** $(.75, .75)$ **and compare your answer to the expected value 1.125.**

(Don't expect exact agreement, because the starting and ending points always are taken to the nearest grid point.)

- **Evaluate the line integral of this field around various closed paths.** Note that as expected, the values of the line integral are non-zero in this case.

A.2   Walk Through: GAUSS Program

The initial screen comes up with the electric field and potential computed for the default charge distribution, i.e., a uniform sphere of charge of radius two units. The three color-coded curves show *E*, *V*, and rho. The dot plot at the bottom shows a slice through one quadrant of the charge distribution.

- **Select Cylindrical and then Planar under the Symmetry menu.** Note the change in the appearance of the three curves, which are recomputed for each symmetry.

- **Select Charge Density Function under the Input menu.**

- **With Symmetry set to Planar, enter the expression** $h(r - 2) * h(5 - r)$. Since $h(r)$ is the unit step function, this represents a slab of charge extending from $x = 2$ to $x = 5$.

- **Select Comparison Function under the Input menu.**

- **In an attempt to match the computed electric field for the slab, enter an expression that you think should describe it.** (The correct expression is given at the bottom of this page if you give up.) Note that after you enter your expression, *x*'s will appear on the plot corresponding to the function you just entered. If your entry was correct, they should lie on the electric field curve. You may also wish to try to guess the form of the potential curve.

- **Reset Symmetry back to spherical.**

- **Click mouse on F2-3D Plots hot key.**

- **Select Potential Function under the Input menu.**

- **Enter the expression 1/r.** Note that the result is a delta-function charge distribution in the 3-D plot.*

A.3   Walk Through: POISSON Program

There is no default system created on the initial screen. In order to create a system, you need to click the mouse on one of the six icons in the "plates" area.

- **Click on the icon that resembles a parallel plate capacitor.** Its voltage will be whatever value is set on the vertical slider, the default being 100 V.

---

*The correct expression for the electric field should be

$$(x - 3.5) * h(x - 2) * h(5 - x) + 1.5 * h(x - 5) - 1.5 * h(2 - x).$$

- **Move the mouse to the center of the drawing area, and click and hold the mouse button as you draw the capacitor.** Releasing the button will complete the drawing.

- **Click on the Run hot key to do the calculation,** and the equipotential contours will be displayed, after the convergence criteria has been satisfied.

- **To get a more accurate result, select Grid Matrix Size under Control, and increase the number of columns in the grid from 40 to 80.** You may only want to do this on a 486 machine; otherwise the calculation may be too slow.

- **Click on Run to redo the calculation with the finer grid.**

- **Select Field Line Through a Point under the Extras menu, and click the mouse at various points to build up a field line pattern.**

- **Click on Esc-Exit to quit this mode.**

- **Select Above Four Plots, under the Plot How menu.** Note the resemblance of the 3-D plot of the potential to a rubber sheet stretched on a wire frame. Also, note how the charge is concentrated at the ends of each plate in the charge distribution plot.

- **Select Contour Map of Potential under the Plot How menu.**

- **Select Cross Section of E and V under the Extras menu.**

- **Move the mouse to various points to see how the listed quantities change.** Note that as expected, Div E is only appreciably non-zero when the mouse is located on the two plates.

- **Click and drag the mouse along a vertical cut across the center of the capacitor,** and observe the plots created when you release the mouse button. Do the same along a cut along the x-axis midway between the capacitor plates.

- **Click on Esc-Exit to return to the original screen.**

- **Click on Del-Delete to start a new configuration.**

- **Click on the icon resembling a row of point charges, and place the charges in the middle of the drawing area by mouse.**

- **When the input screen for the number of charges in the row comes up select the maximum.**

- **Click on Run to calculate the equipotentials.** Note that the row of charges is not an equipotential contour.

- **Select Above Four Plots under Plot How.** Note how the potential varies along the row of charges in the **Potential Plot.** Also, note how the charge distribution (computed from $\nabla^2 V$) is a constant along the row.

- **Select Contour Map of Potential under the Plot How menu.**

- **Delete this system.**

- **Click on the circle icon (a cylinder).**

- **Select Ask for Potential or Charge Function under the Control menu.**

- **Draw a circular cylinder at the center of the drawing area.**

- **When the input screen comes up enter the expression cos(2*pi*t).** This means you have defined the potential on the surface of the cylinder to vary with angle, according to $V = \cos\theta$.

- **Click on Run.** The resulting equipotential contours inside the cylinder should ideally come out to be equally spaced vertical contours (a uniform horizontal field). The fact that they don't is due in part to the finite grid size, and in part to inadequate convergence criteria.

- **Select Relaxation Parameter under the Control menu,** and raise the number of iterations to 300. Note the great improvement in the straightness of the equipotential contours when the calculation is rerun.

A.4   Walk Through: IMAG&MUL Program

The program initiates in the **Image Charge** mode, displaying the simplest situation to which the image charge method may be applied, a single charge near a conducting plane.

- **To display the electric field choose Field Lines—in vacuo only under the Display menu heading.** Field lines are drawn outside the conductor only. Note that they are incorrect since they do not all impinge on the conducting surface in a direction perpendicular to it.

- **Move the close-up window to the upper end of the conducting surface by selecting Move Close-up Window under the Tools menu heading and following the on-screen directions.** This highlights the problems with the current situation since it shows that the intersection angle of the field line with the surface in that vicinity is something like 25 degrees.

- **Now add an appropriate image charge in order to obtain a physically reasonable solution as follows:** Double-click with the mouse at the point at which the image charge should be located. The charge placed at that point will have the present slider value. Using the slider, change the charge to the appropriate value.* If the solution is correct, the field lines will contact the conducting surface everywhere perpendicular to it. The close-up window should show an intersection angle of approximately 90 degrees.

- **Select Field Lines—all space under the menu heading Display to see what the field lines would look like if the conductor was absent and only the two charges were present.** Return to the **in vacuo only** choice under the **Display** menu heading.

---

*The correct solution is an equal and opposite charge located symmetrically within the conducting region.

- **Look at other posed problems. Choose Spherical under menu heading Surfaces.** This problem is solved in Jackson, for example. Note that the image charge is not of the same magnitude as the real charge in this case.

- **To see a case for which a finite image charge solution does not exist choose 120 degree angle under the Surfaces menu heading.**

- **To toggle to the multipole moment mode choose Multipole expansion under the Parts menu heading.** The initial display is of a cylinder (whose cross section in the screen is a rectangle). Superimposed in color is the contour plot of a potential function (whose contours do not fit the cylinder in the slightest). The object is to get better agreement between the cylindrical surface and a contour of the function. Then the exterior problem will have been approximately solved.

- **Using Change moments—keyboard under the Tools menu heading, eliminate the dipole contribution and add the monopole term with amplitude 4.0.** Clearly the cylinder is even with respect to reflection in the x-y plane, and therefore the dipole potential with which the program began was inappropriate. All terms added should be even.

- **Add some quadrupole amplitude by clicking on the appropriate radio button and using the slider to adjust the amplitude.** Investigate additions of either sign and try to decide based on the fit to the cylinder outline whether to add a positive or negative term*.

- **Repeat the above with the hexadecapole (16) contribution to improve fit further.**

- **Select Field Probe under the Tools menu heading and use it to investigate the character of the $\vec{E}$-field associated with this potential function.** There are on-screen directions for using the field probe.

- **Choose another surface to fit by selecting Split sphere I under the Surfaces menu heading. Use the F4-Clear hot key to clear the last potential function off the screen and return to the default dipole potential.** This surface is positive above the x-y plane and negative below. (This is indicated on the screen using dashed curves alternated red and gray for positive and blue and gray for negative.) It thus has odd symmetry and requires the addition of other odd multipole moments. Try to find the optimal additions.[†]

- **Choose Electric field under the Display menu heading to show the electric field associated with this potential function.**

- **Other surfaces to fit can be looked at under the Surfaces menu heading.**

A.5   Walk Through: DIELECT Program

The program begins in the fixed polarization mode and the initial screen displays a 5×15 slab of material with polarization 10 V/a where a is the (arbitrary) cell

---

*The correction addition is $+4$.

[†]This case can be solved analytically. The appropriate values are: (dipole) $3/2$; (octupole) $-7/8$; ('32'-pole) $11/16$.

size. The closeup window in the upper right shows an enlargement of the selected cell (the one with the white x in the main display window) with $\vec{P}$ shown in white, $\vec{E}$ in yellow, and $\vec{D}$ in green. (The latter is too small to be seen in the display, but its components are given quantitatively.)

- **Display $\vec{E}$ everywhere by choosing Electric Field under the menu heading Display.** Yellow arrows showing the magnitude and direction of the field are drawn in all cells on the screen. Note that the field inside the polarized material is reversed in direction due to the polarization. If the field arrows are too large or too small adjust the scale factor with the bottom slider.

- **Adjust the magnitude of the external field using the slider in the lower right window. Set the value to 0.** Note that now the field has roughly dipole character.

- **Probe $\vec{E}$, $\vec{P}$, and $\vec{D}$ in various cells (both empty and occupied) by clicking on them.**

- **Select Cylinder under Configs. menu heading** to see another material configuration. Leave the external field value at 0.

- **Check the nature of the resultant field within the cylinder by clicking on various cells.** Note that the magnitude and direction of $\vec{E}$ within the cylinder are approximately constant. (Deviations are due to the discreteness of the grid being used. Outside the field is again roughly dipole in character.)

- **Now switch to the fixed susceptibility mode by choosing Fixed Susc. (iter to convergence) under the Options menu heading.** In this mode the polarization must be solved for self-consistency by the program. Choose **Polarization** under the **Display** menu heading. Note that because the situation has not yet been "solved" the polarization is non-zero even though the external field value is set to 0.

- **Solve for the polarizations by touching the F2-Calculate hot key.** After the self-consistent solution has been completed, a message to that effect appears at the bottom of the screen. Note that the polarizations have decreased to zero everywhere as expected.

- **Now change to display of $\vec{E}$ under the Display menu heading and then adjust the external field (using the slider) to a value of 5.** Note that the field takes on this value everywhere since the polarizations have not yet been solved.

- **Solve for the polarizations by touching the F2-Calculate hot key.** While the solution is proceeding you will see (unless your machine is too fast) the polarizations developing into a uniform state as expected by theorem.

- **Examine the solution in detail by clicking on cells inside and outside the cylinder. (Field information is displayed in the upper right-hand window.)** Note that the magnitude of $\vec{E}$ decreases markedly upon entering the cylinder, but that the change in $\vec{D}$ is much less, as expected.

- **Choose Double Perpendicular Slabs under the Config. menu heading to examine a different configuration.** After solving note that the slabs do influence one another—the polarizations on the inner surfaces differ from those on the outer surfaces.

- **Set the external field to 0 and solve again using hot key F2-Calculate.**

- **Examine the iterative self-consistent solution step-by-step as follows:** Choose **Fixed Susc. (iterate 1 step)** under the **Options** menu heading. Next set the external field to a value of 5 using the slider. Click on a cell of interest to select it for display and choose **Polarization** under menu heading **Display**. Then touch the F2-**Calculate** hot key to perform one step of the solution. Repeat hitting F2 until the values no longer change. Note that the convergence is oscillatory for some cells.

A.6   Walk Through: ATOMPOL Program

The initial screen comes up with a single atom located in the center of the screen and having the polarizability of potassium. (The latter is among the highest in the periodic table.) An external electric field oriented horizontally from left to right, is displayed in yellow. The white arrow in the displayed atom represents its induced dipole moment.

- **Review field display options under Display menu.** Choose Single Dipole Field to see the field due to the atom present. Choose Combined Field to see the total electric field. All fields are shown in yellow.

- **Review polarization cartoon options under Tools menu.** The default display in the upper right corner of the screen is of the charge cloud of the atom with the nucleus shown shifted relative to the charge cloud. The size of the nucleus and its shift are magnified by 10,000 times in this display mode. Select View Typ. Atom (Nuc. Sc.) to see the shift on the scale of the nucleus.

- **Select the menu item Configs. giving the list of programmed configurations available, and select the choice Horizontal line.** The screen will display 11 atoms in a horizontal row, each having the polarization that would be induced by the external field alone.

- **To see the effect of interactions between the polarized atoms solve self-consistently by touching F2-Solve hotkey.** Note that the polarizations increase due to the interactions.

- **Check the solution in detail by selecting choice Probe Atom under menu heading Tools.** Information concerning atom "a" is displayed. To see which atom is "a" select Label Atoms w/ Letters under the Display menu heading. The atom probed is the last selected atom.

- **Select a different atom to probe it. Click twice (slowly, so that the program doesn't interpret this as a double click) on atom "f" near the center of the screen.** (The first click deselects that atom-the color aqua indicates an unselected atom, and the second reselects it making it the last selected atom.) Now choose Probe Atom again to display the polarization of atom "f". Note that because of the interactions, this atom has a larger dipole moment than did an atom at the end of the line.

- **Change the polarizability of all selected atoms by moving the lower slider to a value of 10.** Solve again using the F2-Solve hotkey.

- **Examine the field quantitatively by selecting choice Field Probe under menu heading Tools.** Click at various points on the screen to display field information in the upper right hand window. A blue field arrow showing the magnitude and direction of the field is drawn at the point of the click. Note that the field is reduced somewhat from its applied value of 5 in the vicinity of the row of atoms, and that the reduction is larger near the center of the row than near the end.

- **Choose Close-Packed Lattice under the Configs. menu heading and solve for the polarization using the F2-Solve hotkey.** The screen displays 19 atoms in that crystal structure.

- **Introduce a potassium impurity at the center of the cluster as follows...** Clear all selections using the F4-Clear hotkey. (All atoms colors will turn to aqua.) Select atom "c" in the center of the cluster by clicking on it. (It will become magenta in color.) Adjust its polarizability to 34 using the slider.

- **Solve again using the F2-Solve hotkey,** and note the distortions of the polarizations of the atoms near the impurity.

A.7   Walk Through: MAGSTAT Program

The program begins by showing $\vec{H}$ for the case of a permanent magnet in the shape of a torus. In this and other displays, the field is shown as a grid of colored arrows.

- **Use the Choose menu to show $\vec{B}$ and $\vec{M}$.** Note that $\vec{B}$ and $\vec{M}$ can be computed and displayed rapidly once $\vec{H}$ is known. Finding $\vec{H}$ at each point is computationally expensive, and must be done initially each time the shape, size, aspect ratio, or permittivity changes. Use the **Switch Units** option under the **Data** menu if you prefer Gaussian units to the SI units that are the defaults.

- **Use the Tools menu to plot graphs of $\vec{B}$ and $\vec{H}$ along some lines that intersect the boundary of the solid.** The algorithm the program uses to compute $\vec{H}$ fails very close to the boundary, so ignore the occasional jagged edge that happens there. Note that the magnitude of $\vec{B}$ varies continuously across the boundary, where it is perpendicular to it, but $\vec{H}$ behaves differently. Use the **Clear** hot key when the screen becomes too cluttered.

- **Use the Display menu to examine the field pattern in three different ways.** Note that the **Tools** menu cannot be used when the magnitude of the field is displayed as a 3-D plot.

- **Use the Shape menu to recompute the fields for a magnet of different shape.** All shapes are solids of revolution around the vertical axis. The

Shape menu does not allow you to change the aspect ratio or size of the solid, which can be done through the **Data** menu. Note especially how the plots of magnetic charge and bound current (see the **Tools** menu) change as the shape changes. If you are impatient with the time required to compute the field, the computation can be speeded up by choosing a coarser grid size under Display option of the **Data** menu.

- **Use the Simulation option under the Data menu to generate the fields around a diamagnet in an external field.** The permeability of the solid is adjusted via a data input screen; diamagnets have permeability $\mu < 1$. The computation of the $\vec{H}$ in the case of a magnetizable material in an external field requires a relaxation algorithm. Repeat for various values of the permittivity and note the changes in the distribution of bound current and magnetic charge.

A.8   Walk Through: ACCELQ Program

The program starts by generating a "snapshot" of the $\vec{E}$-field pattern produced by a point charge in circular motion. The actual motion of the charge is shown in animation at the lower left of the screen.

- **Click in order on each of the choices under the Trajectory menu.** The trajectories increase in complexity from simple 1-D acceleration and deceleration to more complex 2-D motion. Note that it is the change in either speed or direction of charge motion that produces the bending of electric field lines that characterizes a radiation field.

- **Use the arrow hot keys to change the speed of the charge and the time of observation.** The effect of the different keys is shown at the bottom of the screen. Changing the time in increments with the right-angle turn or either 1-D trajectory will allow you to see how a radiation pulse created by acceleration of the charge expands spherically from the point of acceleration.

  If the animation of the trajectory gets tiresome, it can be turned off by an option under the **Data** menu.

- **Click on one of the options under the Explore menu to make the mouse a field probe.** This option allows the user to investigate the relationships between the various fields at various points in the viewing plane. The fields are color-coded, and arrows with lengths proportional to field magnitude are used to represent vectors.

- **Use options under the Plot What menu to generate graphs of various fields as functions of either space or time.** The mouse can be used to specify a line along which the field will be plotted or to specify a point at which the time dependence of the field will be plotted. For example, use the sinuous charge trajectory to explore the relativistic Doppler effect by looking at the time dependence of the magnetic field along points ahead of and behind the charge.

A.9   Walk Through: QANIMATE Program

The default case when the program starts is "turning on," which produces a constant rate of change of direction of motion. If the user presses no keys, this motion will continue and the result will be an animation of the electric field for a circular charge trajectory.

- **Press the space bar, wait a few frames, and press it again. Repeat this process a few times.** The space bar toggles the "turning" on and off, which is a simple way for a user to guide the charge along paths composed of circular arcs and straight line segments. If the charge leaves the viewing screen, it re-enters at another point moving at the same speed and in the same direction, but the accompanying field lines must be regenerated.

- **Press the up and down arrows a few times each.** Note that each press of the arrow changes the speed by a small amount. This tangential acceleration generates a small "kink" that travels outward along each field line. The value of the current speed is shown at the upper left part of the screen.

- **Press the left and right arrows a few times each.** Note that these change the turning rate. If the turning is off, pressing one of these keys will turn it on.

- **Press the function keys F4 and F5 a few times.** Note that these keys do not affect the physical velocity of the charge, but they vary the speed of the animation by changing the rate at which new points on field lines are generated. They also affect the smoothness of the plotted field lines by changing the lengths of the straight segments that make them up. Smooth field lines require more computation between frames, which slows the animation.

- **Press the Esc key to access the pull-down menu.** A data input screen available from the menu provides several display options, which include the ability to plot the spherical "wavefronts" associated with the field pattern.

A.10   Walk Through: EMWAVE Program

The program begins by showing the animation of a plane polarized wave.

- **Use the arrow keys to view the wavetrain from different points of view.** These keys change the altitude and azimuth angles of the viewpoint relative to the reference frame plotted with the wave. The wave is assumed to be traveling in the $+x$ direction.

- **Use hot keys F2 and F3 to change the animation speed.** These keys vary the time interval between animation frames.

- **Change the polarization state of the wave.** The quick and easy way to do this is to choose the circular polarization option under the **Change** menu. Alternatively, you may want to reset the Stokes parameters individually using the **Stokes Parameters** option under **Change** to produce a variety of wave polarizations.

- **Observe the effect on the wave of reflection and transmission at a material interface.** Do this by choosing the **Wave on Interface** option under the **Show What** menu. Press **Enter** to accept the default values for permittivity, permeability, and conductivity to simulate the case of partial reflection and partial transmission into a glossy material. Note the partially standing wave that results from the superposition of incident and reflected waves. Try large values of the conductivity and a circularly polarized incident wave to examine a standing, circularly polarized wave.

- **Choose the Wave + Waveplate option under the Show What menu.** The default configuration shows the conversion of the wave from plane-polarized to circular-polarized. The hot key labeled **Rotate** allows you to see what happens for other orientations of the waveplate axis relative to the direction of plane polarization. A quick **Change** to circular polarization of the incident radiation can be made from the menu to show the reverse process.

- **Try other options under the Show What menu.**

Appendix B

Visualization of Electromagnetic Fields

Because they deal with abstract, three-dimensional (3-D) fields that are sometimes very difficult to visualize, electricity and magnetism can be conceptually rather difficult. Consider, for example, the magnetic field near a bar magnet. We can look at a picture of the magnet itself and get a sense of how it would look and feel if we were actually holding it in our hands. The magnetic field is arguably just as real as the magnet, but we cannot touch it or see it directly. We can see its effects on iron filings, Hall probes, and compass needles; we can describe it mathematically; and we can use words like "lines," "flux," and "permeability" to make the concept less abstract. Still, visualization—actually picturing the field in our minds—is often difficult.

Fast computers with high-resolution graphics give us the ability to turn the abstract mathematics often used to describe electromagnetic fields into graphs and pictures, so computers have great potential for allowing us to visualize fields more easily. However, even computers have their limitations: Like the page of a textbook, the computer screen is two-dimensional (2-D), and drawing a fully 3-D field-line pattern on any such 2-D medium will usually produce a confusing jumble of criss-crossing lines. Each of the simulations accompanying this book uses the computer screen to portray electromagnetic fields, and the authors of the simulations have used a variety of methods for displaying them. Here are some of the techniques we have used:

Two-Dimensional examples. All electromagnetic fields are inherently 3-D, but there are physical systems where it is possible to disregard one of the three dimensions. For example, a very long, current-carrying wire, a long line of electric charge, or slab of material much thinner than its length or width can be thought of as extending to infinity, in which case it is only necessary to show the fields in a perpendicular plane. A system may have cylindrical symmetry, so that it is sufficient to show the fields in just a single plane containing the axis of symmetry. There are other special cases where the electromagnetic fields at all points in a plane are directed either parallel or perpendicular to that plane. In that case, the fields in the plane can be represented without confusion. In all such cases, the users of these computer simulations should remain aware that the screen is portraying only a 2-D slice of 3-D space.

Contours and 3-D Surfaces. Visualization of a scalar fields in two dimensions can be greatly assisted by a contour map in which lines or regions of equal field value have the same color. Although such contour maps aren't as useful in portraying vector fields, the intensity (magnitude) of a vector field is a scalar field that can be represented this same way. A scalar field in two dimensions can also be represented as a 3-D surface, with the value of the field plotted as the third dimension. The CUPS utilities, described in the next section, include both a routine for generating contour maps and one for drawing such 3-D surfaces.

Field-Line Patterns. Vector fields are sometimes shown on the computer screen in these simulations, as they are in textbook illustrations, by drawing continuous field lines. When the field lines are drawn completely and properly, the resulting pattern of lines simultaneously shows both direction and magnitude: The magnitude of the field at each point is represented by the density of lines, and the direction of the field is represented by the tangent to the field line through that point. This method may be the best way to visualize a field in 2-D examples, but it isn't always feasible because of limitations of computing speed, and because satisfactory algorithms for choosing representative field lines are often very difficult to find.

Arrow Grids. A vector field in a plane can be represented by a 2-D array of vectors—arrows whose directions represent the field direction at the grid points. Most textbooks use arrows to represent vectors, with the origin indicating the location of the field point and the length of the arrow representing magnitude. When a grid is formed by many such overlapping arrows, portraying vectors this way can distort the overall field pattern. For that reason we often position the center of the arrow, rather than its origin, on the field point. Similarly, drawing a grid of arrows of varying length can sometimes lead to confusing, overlapping patterns. We have sometimes found it helpful to use color to represent field intensity and to use the arrows only to indicate direction, as if there were a grid of small compass needles mapping the field.

Perspective Projection. Sometimes it is necessary to visualize fully 3-D pictures of fields for which none of the above techniques are adequate. In that case, 3-D visualization can sometimes be accomplished by viewing in perspective projection from more than one viewpoint. While binocular viewing hasn't proved feasible in these simulations, it is possible to portray a field pattern or a 3-D surface by viewing it in perspective projection from a variety of individual viewpoints. When the viewer has control over the viewpoint, the illusion is similar to that of being able to examine the shape of a 3-D object by turning it in the hands.

Appendix C

Units of Electromagnetic Quantities

The proper use and understanding of units and dimensions is important in every field of science and every subfield of physics. However, units seem especially confusing in the study of electricity and magnetism. Units for physical quantities chosen for convenience in particular applications at some period in history aren't always the most convenient for the same quantities used for broader application at later times. This is especially true for electricity and magnetism, which has an extremely broad range of applicability, and where units in use today were often chosen before the connection between electric phenomena and magnetic phenomena was fully appreciated.

Two major systems of units are in general use in science and engineering fields that involve electricity and magnetism: the SI system of units (Systeme International d'Unites) and the Gaussian system. The two systems are often simply called "mks units" or "cgs units," respectively. Even the dimensions assigned to various electromagnetic quantities are different in these two systems, and so are the forms of many of the equations that represent the fundamental laws of electricity and magnetism.

SI units are rationalized mks units, or more properly MKSA (meter-kilogram-second-ampere) units. In SI, electric current is considered a fourth fundamental dimension beyond the three of length, mass, and time used in mechanics. The SI system is advantageous for engineering and similar practical fields of application, and it has been adopted in most recent textbooks used in undergraduate physics. Therefore, this book also uses SI (mks) units.

The Gaussian system of units is a cgs (centimeter-gram-second) system in which there are only three fundamental dimensions, because electric charge is defined in terms of length, mass, and time by setting the constant in Coulomb's law equal to unity. That is, one Gaussian unit of charge (the esu) is defined to be that amount of electric charge that exerts a force of 1 dyne on the same amount of charge 1 centimeter away. The units of all electromagnetic quantities in the Gaussian system can be expressed in terms of just three arbitrarily chosen units—the centimeter, the gram, and the second.

The Gaussian system has the advantage that the basic equations of electricity and magnetism contain the speed of light c as the only fundamental universal constant, instead of the ϵ_o and μ_o that are ubiquitous in SI units. The appearance of c in the equations often clarifies the role of special relativity in the theory of radiation and in the connection between electical phenomena and magnetic phenomena. Further, the dimensions of the commonly used fields $\vec{E}$, $\vec{D}$, $\vec{B}$, and $\vec{H}$ are all the same in Gaussian units. In effect (if not always in name) the units of all four fields are also the same. For those reasons and others, the Gaussian system is commonly used by theoretical physicists and in subfields of physics less closely related to engineering, such as quantum mechanics or astronomy.

Compounding the unit problem is the fact that units from the two sytems are often intermixed in practice. For example, the same physics laboratory that uses meters calibrated in volts, amperes, and ohms will often use magnetic field probes calibrated in gauss, and electromagnets rated in kilogauss. Therefore, just as every good physics student should quickly be able to convert the units of mechanical quantities like force and energy between cgs and mks systems, students of electricity and magnetism should learn to convert units of electromagnetic quantities between SI and Gaussian systems. For example, it is good to remember that the charge on a proton is 1.6×10^{-19} coulombs only when using SI units—it is 4.8×10^{-10} esu in the Gaussian system.

Writing computer code for electricity and magnetism presents yet another facet of the units problem. The bits and bytes of information manipulated by the computer are pure numbers without dimension, but they usually represent the values of dimensioned physical quantities. Overflow or underflow errors can be the programmer's punishment for being careless in the choice of units in writing the code for a given application. It is common that there will be a "natural" system of units to use within the computer code for a given application that will give moderate numerical values to physical quantities, and this natural system may be neither SI nor Gaussian.

Therefore, in some of the simulations accompanying this book both the code and the results of the calculations employ neither SI nor Gaussian unit systems. For example, the simulations dealing with electromagnetic radiation use systems of units where the speed of light is the unit of velocity ($c = 1$). Other simulations use grid spacing as a unit of distance and an arbitrary unit of charge so as to simplify or clarify numerical algorithms. While students may find this confusing at first, learning to think of units as tools and to use these tools to good advantage is essential training for any scientist or engineer.

Index